The Future of the Curriculum

D0465692

This report was made possible by grants from the John D. and Catherine T. MacArthur Foundation in connection with its grant making initiative on Digital Media and Learning. For more information on the initiative visit http://www.macfound.org.

The John D. and Catherine T. MacArthur Foundation Reports on Digital Media and Learning

Peer Participation and Software: What Mozilla Has to Teach Government by David R. Booth

The Future of Thinking: Learning Institutions in a Digital Age by Cathy N. Davidson and David Theo Goldberg with the assistance of Zoë Marie Jones

Kids and Credibility: An Empirical Examination of Youth, Digital Media Use, and Information Credibility by Andrew J. Flanagin and Miriam Metzger with Ethan Hartsell, Alex Markov, Ryan Medders, Rebekah Pure, and Elisia Choi

New Digital Media and Learning as an Emerging Area and "Worked Examples" as One Way Forward by James Paul Gee

Digital Media and Technology in Afterschool Programs, Libraries, and Museums by Becky Herr-Stephenson, Diana Rhoten, Dan Perkel, and Christo Sims with contributions from Anne Balsamo, Maura Klosterman, and Susana Smith Bautista

Young People, Ethics, and the New Digital Media: A Synthesis from the GoodPlay Project by Carrie James with Katie Davis, Andrea Flores, John M. Francis, Lindsay Pettingill, Margaret Rundle, and Howard Gardner

Confronting the Challenges of Participatory Culture: Media Education for the 21st Century by Henry Jenkins (P.I.) with Ravi Purushotma, Margaret Weigel, Katie Clinton, and Alice J. Robison

The Civic Potential of Video Games by Joseph Kahne, Ellen Middaugh, and Chris Evans

Quest to Learn: Developing the School for Digital Kids by Katie Salen, Robert Torres, Loretta Wolozin, Rebecca Rufo-Tepper, and Arana Shapiro

Measuring What Matters Most: Choice-Based Assessments for the Digital Age

by Daniel L. Schwartz and Dylan Arena *Learning at Not-School? A Review of Study, Theory, and Advocacy for Education in Non-Formal Settings* by Julian Sefton-Green

Stealth Assessment: Measuring and Supporting Learning in Games by Valerie Shute and Matthew Ventura

The Future of the Curriculum: School Knowledge in the Digital Age by Ben Williamson

For a complete list of titles in this series, see http://mitpress.mit.edu/books/series/john-d-and-catherine-t-macarthur-foundation-reports-digital-media-and-learning.

The Future of the Curriculum

School Knowledge in the Digital Age

Ben Williamson

The MIT Press
Cambridge, Massachusetts
London, England

MIT Press books may be purchased at special quantity discounts for business or sales promotional use. For information, please email special_sales@mitpress.mit.edu or write to Special Sales Department, The MIT Press, 55 Hayward Street, Cambridge, MA 02142.

This book was set in Stone Sans and Stone Serif by the MIT Press. Printed and bound in the United States of America.

Library of Congress Cataloging-in-Publication Data

Williamson, Ben (Educator).
The future of the curriculum : school knowledge in the digital age / Ben Williamson.
    pages   cm.—(The John D. and Catherine T. MacArthur Foundation reports on digital media and learning)
Includes bibliographical references.
ISBN 978-0-262-51882-6 (pbk. : alk. paper) 1. Education—Curricula. 2. Curriculum planning. 3. Education—Effect of technological innovations on. 4. Digital media. I. Title.
LB1570.W5765 2013   375'.001—dc23   2012038069

10 9 8 7 6 5 4 3 2

Contents

Series Foreword   vii

1   Introduction: Prototyping and Researching the Curriculum of the Digital Age   1

2   Curriculum Change and the Future of Official Knowledge   15

3   Networks, Decentered Systems, and Open Educational Futures   31

4   Creative Schooling and the Crossover Future of the Economy   47

5   Psychotechnical Schools and the Future of Educational Expertise   65

6   Globalizing Cultures of Lifelong Learning   85

7   Making Up DIY Learner Identities   101

8   Conclusion: An (Un)official Curriculum of the Future?   115

Notes   125

Series Foreword

The John D. and Catherine T. MacArthur Foundation Reports on Digital Media and Learning, published by the MIT Press in collaboration with the Monterey Institute for Technology and Education (MITE), present findings from current research on how young people learn, play, socialize, and participate in civic life. The Reports result from research projects funded by the MacArthur Foundation as part of its $50 million initiative in digital media and learning. They are published openly online (as well as in print) in order to support broad dissemination and to stimulate further research in the field.

1 Introduction: Prototyping and Researching the Curriculum of the Digital Age

Digital media and learning has become a critical area for educational research in the twenty-first century. Yet little research has been carried out on the practical and conceptual implications for the school curriculum in the digital age. This report asks a very simple question: what might be the future of the curriculum in the digital age? It examines a series of twenty-first century curriculum innovations in order to show how various ideas about the future curriculum are now being styled into school practice, and it seeks to understand the emerging issues raised by meshing the curriculum and digital media together.[1] It explores a range of contemporary social, political, economic, and cultural issues facing the future of the curriculum and examines the production of ideas about the practical organization and planning of a future curriculum. What kinds of visions for the curriculum of the future are being imagined, invented, and promoted? The main argument is that any curriculum always represents a certain way of understanding the past while also promoting a particular vision of the future. To use pragmatist philosopher William James's metaphor, the curriculum is a "saddleback" with both a rearward-looking and a forward-looking trajectory. It expresses simultaneously a legacy from the past and aspirations and anxieties about the future.

The case studies are a selection from a growing number of curriculum innovations that correspond with a new globalized era of networked technologies, communications, and digital media. They originate from the United States, the United Kingdom, and Australia, and they involve a variety of actors and agencies from the public, private, and philanthropic and nonprofit sectors. These programs act as micro-level sites of curriculum reform that refract macro-level ideas about social and technological transformation. The analysis asks what these curriculum prototypes select from the past, how they represent the present, and what ideas they generate about the future. Collectively, they represent a new "style of thought" about the school curriculum for the digital age.

In light of the aspirations and objectives of these programs, what could the curriculum of the future look like? What knowledge should it contain? What visions of the future do these curricular prototypes promote and catalyze? What individuals and organizations are involved in designing and promoting them, and on what expertise and authority? What wider social, cultural, economic, and political associations and objectives are embedded in them? And, most important of all, how do such curricula seek to shape the minds, mentalities, identities, and actions of the young?

Microcosmic Futures

The curriculum is a microcosm of the wider society outside school. It constitutes what a society elects to remember about its past, what it believes about its present, and what it hopes and desires for the future. It is both retrospective and prospective, and it encourages learners to look back at the past and look forward to the future in particular ways. The design of a curriculum shapes the minds and mentalities of young people and encourages them to understand and act in society in particular approved ways. As a result, the local detail of all curriculum reform needs to be understood and grounded in long waves of societal change that are pursued from the past into the present and from there projected into the future.[2]

Understanding curriculum reform in this way alerts us to how major reform movements and policies such as A Nation at Risk and No Child Left Behind have been assembled through debates, conflicts, and political activities that have themselves been shaped through other social and historical events, and that have led to the production of normative visions of the future. In fact, it was A Nation at Risk that, during the Reagan administration in 1983, argued the case for educational reform on the basis that "knowledge, learning, information, and skilled intelligence are the new raw materials of international commerce" and "the indispensable investment required for success in the 'information age' we are entering." A Nation at Risk presented long waves of change—in the form of the globalization of commerce in an "information age"—as the context for the promotion of a future "Learning Society" that was to be extended into the local details of the traditional institutions of learning, schools and colleges, and beyond them into the microlocalities of "homes and workplaces; into libraries, art galleries, museums, and science centers; indeed, into every place where the individual can develop and mature in work and life." Since the early 1980s, then, educational and curricular reforms have been widely premised on the perceived incapacity of schools to keep pace with technological change and its social and economic implications. Much of this argument remains familiar in talk of digital age reforms some thirty years later, as we continue to ride the crest of a long wave of educational change.[3]

All of the curriculum prototypes examined in this report offer a view of how the curriculum might be redesigned and reformed in the perceived context of the digital age. They all start with the same basic assumption that new and constantly changing technologies, accompanied by complex, long waves of social and technological change in the economic, political, and cultural dimensions of existence, have contributed to the need for curriculum reform. These assumptions are part of an emerging "style of thinking" about modern society. The dominant style of thinking about society in today's digital age is saturated with "cybernetic" metaphors of information, networks, nodes, dynamics, flexibility, multiplicity, speed, virtuality, and simulation. This is not to say that we live in cybernetic societies, but in societies that are increasingly understood and consequently shaped through a cybernetic style of thought. A style of thought is a particular way of thinking, seeing, and practicing. It designates what counts as an argument or an explanation in a particular field, underpinned by key terms, concepts, references, relations, and techniques of intervention. But it doesn't only explain: it actually shapes and establishes the problems, difficulties, and issues for which an explanation is required. Rather than being solely explanatory, then, a style of thought modifies or remakes the very things it explains.[4]

The trend in curriculum making examined in this report is therefore far from a neutral or nonpolitical activity: it involves a cybernetic style of thought that pervades attempts both to explain and to remake the links between curriculum and society in the digital age. The curriculum of the future is not "out there" waiting to be discovered, but must be imagined and constructed. It is important to treat these programs and their objectives not simply as microcosms of a world that already exists, but as microcosms of imagined futures being prefiguratively practiced, or microcosmic futures still in the making.[5]

Because aspirations for the curriculum are linked together with the global concerns of the digital age, the future of the curriculum has become a subject of intense debate. Perhaps more than any other aspect of schooling, new technology and digital media are matters of significant interest for a wide range of parties that extend beyond the formal organs of education systems. For example, almost all of the transnational computing companies have significant educational programs and funding initiatives. Microsoft, Google, Mozilla, Apple, Cisco, Hewlett Packard, and so on have all made high-profile statements about the need for schools to keep pace with technological advances. Commercial participation in curriculum design and research is now a serious matter for research.[6]

Besides governmental and commercial interests, many philanthropic organizations, foundations, charities, and nongovernmental and nonprofit organizations have also put digital media and learning at the heart of their operations. Political think tanks, pressure groups, and semi-governmental agencies too have attempted to prioritize technology on the educational policy agenda. Supranational and multilateral bodies such as the Organisation for Economic Co-operation and Development (OECD), the United Nations (UN), the World Bank, and the United Nations Educational, Scientific and Cultural Organization (UNESCO) have all made recommendations and specifications for educational programs. All of this is evidence of a transformation in how the job of public education gets done—increasingly, by third parties doing parts of its work from within. More than ever, curriculum planning is being performed in an "unreal world" at a distance from the day-to-day tasks of schools.[7]

Additionally, many of today's digital kids seem to recognize the problem of the content curriculum, standardized testing, and credentialing just as well as many critical curriculum scholars, digital media researchers, and global Internet entrepreneurs do. According to some optimistic accounts, young people today are sophisticated cultural producers of digital media, actively creating, remixing, and circulating content online in complex ways that far outstrip anything demanded of them by the traditional subject curriculum. More critical analyses suggest that they are being lured by a seductive commercial curriculum and public pedagogies of advertising into cultures of consumerism and materialism. Taking a more balanced view, digital media, as an important part of young people's lives and cultural experiences, offer forms of participation, community, belonging, and communication that are important and meaningful; at the same time, the meanings that may be derived by young people are subtly shaped and limited by consumer culture.[8]

The task of reforming the curriculum of the future, then, is a matter of political change in education systems as well as a matter of changing what teachers and children do in schools. Curriculum reform changes the nature and structure of the connections between various political centers and nonpolitical authorities and the distant microlocalities of educational practice and experience.[9] The case studies discussed in this report are the products of a variety of surprising alliances between actors and agencies from well beyond the confines of traditional government bureaucracies and education systems, and from a variety of intellectual sources rather than from any single political perspective, academic orientation, or particular ideological position. In this synthesis and juxtaposition of agents and agencies, all sorts of arguments, rationales, and objectives for the curriculum are bundled up and packaged together. The curriculum prototypes examined are examples of an increasingly globalized educational reform network within which new educational ideas, trends, and fashions are being borrowed, copied, interconnected, harmonized, and hybridized across distant and local sites.[10]

"Centrifugal schooling" is the collective name used in this report for the prototypical curricula of the future emerging from these networks. The projects are each distinctive and innovative in their own unique ways, yet they share similar concerns, identify similar problems, and propose similar solutions.[11] Centrifugal schooling expresses a vision of the future of education and learning that is decentered, distributed, and dispersed rather than narrowly centered, channeled, and canalized. Its keywords are "networks," "connections," and "decentralization," as well as a family of related centrifugal terms. These keywords articulate a shift from a centered tradition of thinking about schooling, as an institutional process that happens on school premises through formal pedagogic techniques of transmission, to an emerging decentered vision where learning is centrifugally dispersed and cybernetically distributed into society through new technologies, communication networks, the informal pedagogies of media, and emerging social practices of interest-based, peer-to-peer, just-in-time participatory learning.[12] These ways of thinking about twenty-first-century learning are related to the general sense that social reality today is less securely anchored or embedded in the traditional institutions that patterned social, cultural, and personal life in the past—namely, families, social classes, religious affiliations, lifelong vocations, and so forth. Instead, our social structures and institutions today are more scattered, fluid, disorganized, disembedded, diverse, mediated, risky, individualized, and confusing.[13] Networked communication technologies are fast becoming part of this mobile social environment. Internet users are no longer configured as the recipients of unidirectional flows of broadcast material generated from centers of media production but as multidirectional nodes in complex convergent communication circuits and network flows.[14]

Recast as a response to these technological changes, the kind of prototypical curriculum of the future associated with centrifugal models of schooling may be imagined as a more "open source" process rather than a fixed product, as embodied in the "wiki" format of open authorship, collective editing, and collaborative production. Crudely caricatured, the traditional centered curriculum was a curriculum based on a standardized mass-production model of "reading" that positioned teachers as broadcasters and learners as receivers, as embodied by school textbooks. In comparison, the decentered curriculum is a post-standardized, mass-customizable "read-and-write" curriculum that repositions teachers and learners as peer-to-peer producers, participative authors, and active creators of curriculum content, processes, and outcomes in a distributed meshwork of joined-up learning. A "wikiworld" of new learning encompasses a move away from seeing curriculum as a core canon or central body of content to seeing curriculum as hyperlinked with networked digital media, popular cultures, and everyday interactions.[15] Consequently, it is now becoming possible to conceive of the future of schooling itself as a network-based distributed system of learning rather than a strictly routinized series of teaching tasks, though there is little evidence of the institutionalization of these methods.[16] That lack of evidence so far makes the research on the future of the curriculum for the digital age all the more significant. Furthermore, such styles of thinking about the future of learning are not all new and historically unique, as shown by the surprising continuities between politically conservative policies like A Nation at Risk, with its calls for a "Learning Society," and more recent advocates for "24/7 learning everywhere."[17] Centrifugal schooling is also continuous with a "connectivist" style of curriculum thought that was popularized in the 1990s, which today is being updated and projected into a hyper-connected "network" future. The changes embodied by centrifugal schooling are gradual, incremental, and cumulative, rather than representing an epochal break with the past.[18]

Researching Curriculum Networks

This research follows critical curriculum scholars in exploring two perspectives. First, from a critical theory perspective, it asks how the curriculum of the future may reflect the social power, interests, politics, and ideologies of particular groups in society. What different purposes and views of the future of society do they deploy, and how are these embedded in their curriculum concepts? Second, however, the analysis takes up a more "poststructuralist" view that social power does not emanate from a single dominant ideological source that produces the curriculum, but that the curriculum is produced within a complex web where power and influence are continually shifting and subject to continuous negotiation. The projects and programs under scrutiny are not big-P policies or official curriculum reforms but little-p policy proposals and reforms-in-action. Consequently, the analysis looks beyond the power, ideology, and influence of the "usual suspects" of government departments and big commerce to trace the "micro-level actors" involved in the reimagining of the curriculum and the norms and values it embodies.[19]

In order to interrogate the curriculum of the future imagined by the prototype projects, this report will examine how curricula are created and distributed through *curriculum texts* and *curriculum networks*.[20]

Curriculum Texts

Curriculum texts are documents that introduce and explain curriculum ideas. They include curricular guidance, research reports, Web sites, resources, and materials provided by the various creators and sponsors of curriculum projects. These texts take ideas about alternative possible future directions for the school curriculum and translate them into proposals for programs and practices. Texts are a useful source of documentary evidence because they render complex ideas coherent and communicable, though for that reason they do need to be read with critical caution as selective representations rather than as empirical observations. All educational texts, as relays of styles of thought, create positions for teachers and children, managers, parents, policymakers, and so forth, providing them with a language, vocabulary, and a repertoire of practices with which to think and act. They make particular sets of ideas, language, vocabulary, and concepts obvious, commonsense, and seemingly true. What such a text analysis approach aims to uncover is the distinctive style of thought regarding the curriculum of the future that runs through these projects—its terms, concepts, references, relations, arguments, and explanations, as well as associated practical techniques for curricular intervention. Texts such as those interrogated in this report are understood to exert and produce real effects, though the extent to which they actually produce what they envision remains a matter for further empirical research.[21]

Curriculum Networks

The research also traces something of the networks of relations between various actors involved in designing the curriculum of the future. The curriculum is understood as assembled and made up through interactions between agents and agencies of many kinds—individual people, parties, organizations, companies, networks, institutions, and so forth—as well as texts, technologies, and objects, rather than predetermined as a complete and coherent product or a black box constituted by a universally given body of knowledge or by predetermined purposes and aims. As a consequence, the approach in this report is to focus on curriculum texts as documentary constructions of reality that are constantly being circulated, moved on, and connected up to other actors and things. A curriculum is actively assembled, improvised, and "lashed up" from a messy and heterogeneous mix of people, groups, coalitions, organizations, institutional structures, each associated with different ideas, theories, and knowledge; political, intellectual, and historical associations; and a panoply of ongoing negotiations, decision making, and compromises. The production of a curriculum for the digital age is embedded in theories of learning and pedagogy, and assumptions about new technology and media that are all imbued with political, cultural, and economic values and objectives. The participation of such diverse players and elements introduces a variety of sources of authority and expertise into the curriculum-making mix. These participants and elements join together as networks, sometimes fleetingly, sometimes for long enough to establish and maintain projects based on a coherent shared vision, occasionally with sufficient durability to achieve something like system-wide influence. Importantly, taking this view forces researchers to consider the ways in which the curriculum may be shaped by actors and forces acting on it "at a distance"—that is, not through direct manipulation or influence but through delicate connections from afar. A curriculum possesses, so to speak, a messy social life. It is the result of myriad local and distant attachments between people and their historical, conceptual, and political networks, and it is assembled according to specific negotiations and compromises concerning which knowledge and legacies from the past and which future visions of a society are to be included or excluded from it.[22]

The Case Studies

The curriculum R&D programs examined include the following:

Enquiring Minds (EM) was a curriculum R&D project carried out over a four-year period between 2005 and 2009 by the nonprofit organization Futurelab in the city of Bristol in the United Kingdom, with funding from Microsoft Partners in Learning. Initially, two schools participated in the trial, with students aged 11–13, though it was later disseminated widely. It aimed to produce an approach to curriculum based on a dynamic view of knowledge and "the challenges schools face in the task of preparing children for a future characterized by rapid social, technological and cultural change."[23]

High Tech High (HTH) was originally launched in 2000 as a single charter school by a coalition of San Diego business leaders.

Built around a project-based curriculum, HTH is intended to "integrate technical and academic education to prepare students for post-secondary education in both high-tech and liberal arts fields." It has since evolved into an integrated network of eleven public charter schools in San Diego County, a teacher certification program, and a new Graduate School of Education, with financial backing from the Amar Foundation, Simon Foundation, and the James Irvine Foundation.[24]

Learning Futures aims to support students to "work and thrive as the world grows more interconnected, the environment becomes less stable, and technology continues to alter relationships to information." Established in 2008 by the nonprofit Innovation Unit and the philanthropic Paul Hamlyn Foundation in London, Learning Futures has worked with forty schools to develop innovative changes to curricula, pedagogy, and assessment. In early 2012 it published a collaborative guide to project-based learning in partnership with High Tech High Graduate School of Education.[25]

New Basics was originally trialed in 2000–2004 in more than fifty schools in Queensland, Australia, with support from the state government department of education. It promoted "futures-oriented categories for organizing curriculum" and a way of "managing the enormous increase in information that is now available as a result of globalization and the rapid change in the economic, social and cultural dimensions of our existence."[26]

Opening Minds, initiated by the Royal Society for Arts, Manufactures and Commerce (RSA) in the United Kingdom as a "competence-based curriculum which aims to equip young people with the skills they will need for life and work in the knowledge-intensive and new media-rich 21st century." Initially trialed for three years (beginning in 1999) in a small cluster of British secondary schools with students aged 11–14, by 2011 the competencies curriculum had extended to a network of 200 schools nationwide, established it own flagship school in Manchester, and become an independent charitable organization.[27]

Quest to Learn (Q2L) is a "school for digital kids" that opened in New York City in 2009. A collaboration between the nonprofit Play Institute and the education reform organization New Visions for Public Schools, the Q2L curriculum and pedagogy emphasize "design, collaboration, and systems thinking as key literacies of the 21st century."[28] A sister school was established in Chicago in 2011.[29] Both receive support and funding from the John T. and Catherine D. MacArthur Foundation.

In addition to these specific programs and schools, the report also looks at two major partnerships:

The Partnership for 21st Century Skills (P21), a national organization in the United States that advocates for "21st century readiness for every student." Its Web site states that: "As the United States continues to compete in a global economy that demands innovation, P21 and its members provide tools and resources to help the U.S. education system keep up by fusing the 3Rs and 4Cs (Critical thinking and problem solving, Communication, Collaboration, and Creativity, and Innovation)." P21 members include many high-tech multinational corporations.[30]

The *Whole Education* alliance in the United Kingdom represents a network of charitable, nonprofit, and other "third sector" educational organizations. Whole Education brings together education organizations that demonstrate "a commitment to developing a range of skills, qualities, and knowledge that young people will need for the future," providing a mix of "practical and theoretical learning," and thereby "recognize that learning takes places in various settings, not just the classroom."[31]

2 Curriculum Change and the Future of Official Knowledge

Understanding the school curriculum has a long intellectual history. Yet the links between curriculum theory and digital media are less well developed. This chapter establishes some important insights from curriculum research for the study of the future of the curriculum in the digital age. The key issues concern what counts as legitimate or official school knowledge and who gets to legitimize it. The questions, then, are what knowledge is to be included in the curriculum of the future, what are its origins in the past and the cultural legacies it represents, what future does it envision, and what authorizes its inclusion?

The chapter introduces some useful concepts for considering curriculum change and provides a brief historical overview of curriculum change over the last three decades. It then describes some contemporary examples of curriculum programs and examines them as microcosmic condensations of current social changes.

Curriculum Change

Curriculum is the intellectual center of schooling and its main message system. It links together academic and vocational knowledge and skills with personal identity and the public culture of society. It states what is to be studied and the modes of inquiry for studying it. At its narrowest a curriculum specifies the content of specific subjects. More broadly it describes the values and aims used to justify the total program of an educational institution and all of the educational processes and learning that go on within it.[1]

Looking at it more politically, the curriculum consists of practices that carry specific meanings and importance in society. The curriculum acts, then, as a conduit for other forces and conflicts in society. It is absorbed in complex social, cultural, political, and economic debates and conflicts concerning who gets to "select" for inclusion what counts as "official knowledge." In some countries, the school curriculum is specified at the national level, as a national curriculum. In the United States, state textbook adoption policies and major federal policies such as No Child Left Behind have been described as a "hidden" national curriculum. It exerts powerful effects on students, structuring the ways in which they comprehend the world they encounter, promoting norms of acceptable conduct in society, and functioning to reproduce political, social, and class structures.[2]

Any efforts to change the curriculum, the epicenter of schooling, can send seismic shockwaves through schools and beyond into society itself. Most curriculum change follows this simple formula:

1. A preferred vision of society is identified.

2. The conditions for the existence of such a society are then identified.

3. The role of the education system and the contents and form that a curriculum should take to achieve these social ends are clarified.

4. The delivery of the means to those ends are then enacted, resulting in changes to existing curricular forms and changes to society.[3]

This formula underpins the process by which most curriculum change is reasoned out, planned, and implemented. Curriculum change, therefore, is a political act, motivated by particular interpretations of educational purpose, aspirations for the future, and ideas about the kinds of people that a society expects to emerge from school. Often curriculum reform depends on the manufacturing of educational crises, disinformation, myths, and half-truths. The educational status quo is attacked in order to bring about a different, seemingly better future. A curriculum, then, represents a particular representation of reality and constitutes a set of messages about the future. It represents what counts as "official knowledge."[4]

Factory Schooling

Since the 1980s, official educational reform in the United States, the United Kingdom, Australia, and New Zealand, as well as elsewhere, has been driven by a very particular preferred vision of society. The vision is of a high-tech, global high-skills economy, with education geared to enhancing competitiveness.[5] If the curriculum of the past could be characterized as "factory schooling"—with the great mass of students working on an assembly line of facts and tests in preparation for life in largely routine low-skills industrial jobs—then the curriculum of the future for 1980s curriculum planners was to be focused on the production of a more educated, flexible, and highly skilled workforce.[6] The factory schooling model had become untenable because the factory had been eliminated as a source of employment.

In the United States this future vision was first articulated by the 1983 policy report *A Nation at Risk: The Imperative for Educational Reform*. The report put US public schools under a concerted siege of reform strategies organized around the discourse of competition. It articulated the conservation of Western values and knowledge through a future economic vision of enterprise and entrepreneurship to be taught in "high-tech" schools. Meanwhile, in the United Kingdom a "Great Debate" on the perceived need to link the curriculum to the needs of industry led directly to the establishment of the National Curriculum. On both sides of the Atlantic, this period saw a gradual merging of both the economic and cultural dimensions of the curriculum. Economically, the curriculum was now to be modernized in order to ensure global competitiveness in a free market; culturally, it was intended to protect Western values and knowledge, or to conserve culture, in an increasingly globalized context.[7]

The result has been a tendency to see the curriculum as a cold and mechanical product for ensuring economic competitiveness and protecting conservative Western culture, especially business culture. Scholarly studies of curriculum change in the United States since the 1980s have shown how teachers have gradually lost control of curriculum change processes and arguments, while state bureaucracies, corporate organizations, and religious leaders have competed to govern it. Teachers, once positioned as "factory workers," have been reconceived as managers of student learning in institutions that promote business values of outcomes, productivity, the bottom line, accountability, and standards, with the emphasis on delivering individualized, skills-based instructional programs. In the United Kingdom, too, the standardization of curriculum and testing has been criticized for producing "factory schools" and manufacturing learners who are little more than well-drilled automatons.[8]

Flat Learning

The end of factory schooling has been accepted simultaneously by industry, by modernizers, and by radical educators opposed to its narrow economic instrumentalism on progressive humanist grounds. Oddly enough, curriculum reform after the elimination of factory schooling has become a joint enterprise between economic modernizers on the right and radicals on the left who have accepted the basic argument that the reinvigoration of the economy in a "cybernation" depends on the transformation of schooling and the premise that high-skills schooling will be more equitable for all.[9]

In the 1990s the hardline curricular fundamentalism of the conservative restoration came under attack from researchers who described it as promoting a regressive and retrospective "curriculum of the dead" with a structural resemblance to medieval schooling.[10] The emphasis, since the late 1990s, has been on creative and innovative futures that depend on greater curricular flexibility rather than selective rigidity. In this era, it is claimed, knowledge and creativity have higher economic and cultural value than manufacturing or physical products and economic restructuring depends on high-tech innovations in new technology and media. Consequently, greater emphasis is put on education to teach the cognitive skills associated with knowledge work, on the production of ideas, knowledge, and information rather than material "stuff."[11]

The knowledge economy has become the dominant political style of thought in education reform worldwide today.

The knowledge economy is used both as an explanation and as a rationale for the modification of the curriculum. In the knowledge economy style of thought, knowledge is assumed to be at the heart of economic competitiveness. Better educated nations therefore have an advantage in the global economy, while well-educated students can aspire to high status, high-skills knowledge jobs that can in turn assure them of rapid upward social mobility. Portfolio careers without boundaries replace lifelong employment. Muscle power is replaced by brainpower in the search for competitive advantage, and value is derived from integrating behavioral competencies with modular task components. That is, if the global economy is based on increased flexibilization, componentization, and modularization of work, then it will require congruent educational practices—flexible, component-based, modular curricula. It is an imaginary and highly politicized narrative of how the economic world is structured and how individuals, namely students, can play their part in its success. A flat world, so the narrative goes, requires flat education systems, although the evidence that this theory works is seriously debatable and its political conviction in competitive global free trade needs to be treated cautiously.[12]

The result has been a thoroughgoing reimagining of the purposes of education, most spectacularly demonstrated by a massive investment in computing facilities in schools around the world. More subtly but more importantly, questions concerning the curriculum have been pushed aside as the emphasis has been put on skills, competence, thinking, and other categories of learning for the twenty-first century. This is the result of the argument that "know-how" is now more important than "know-what," since most knowledge learned at school—as contained in the curriculum—is likely to become outdated very quickly in a world that is in hyperdrive.

Leading such arguments, researchers from the interdisciplinary field of the "learning sciences" have emerged as a dominant source of authority and expertise on the structure and organization of pedagogy and learning in schools. This interdisciplinary blend of cognitive science, educational psychology, and computer science (and increasingly neuroscience) is intellectually rooted in constructivist, constructionist, sociocognitive, and sociocultural theories of learning rather than in the societal issues that motivate most curriculum research. It emphasizes the design and application of new instructional programs and ICT applications that can "transform the future of learning" across a spectrum of "schools, homes, workplaces and communities."[13] Instead of focusing on the structural question of how formal education is organized and how knowledge is selected and presented for study, learning scientists concentrate on improving learning, on questions of intelligence and thinking, on building learning power, on enhancing cognition and metacognition or "learning how to learn"—all aspects of brainpower. A plethora of frameworks of skills, behavioral competences, and new literacies now compete with one another to better align the education system with contemporary challenges and the curriculum and knowledge have been marginalized as monolithic relics of a former era while the science of twenty-first-century learning and the promotion of brainpower has been established as a new educational common sense. In place of curriculum, a "new language of learning" has been assembled by learning scientists from a composite of constructivist and sociocultural theories of active knowledge construction, increased emphasis on generic learning outcomes, and a psychological view of the learner.[14] Rather

than flat *education* systems, then, we are witnessing the rise of a
flat *learning* system as the science of learning and building brainpower is applied right across the full range of formal and informal situated contexts, both in the real and virtual worlds.

The science of brainpower has been adopted by enthusiasts
for the knowledge economy and the digital age. The hybridization of the learning sciences with the cyberutopia of a knowledge economy suggests that a science of future-building has
been discovered: now that we can transform how people learn,
we can calculate how to construct the future by investing in
brainpower. Educational policies and reform ideas now routinely espouse such a science of future-building.[15]

For curriculum researchers this position raises serious political questions. The focus on a science of learning and learners
in a high-tech computerized knowledge economy deflects attention away from wider social issues and questions about the links
between school and society. The science of skills and know-how
evacuates curricular knowledge of its authority and replaces the
terms "education," "school," and "curriculum" with "learning," "learning styles," and "learning centers." The result, oddly
enough in a knowledge economy or a knowledge society, is that
knowledge seems to lose all its authority and the curriculum is
emptied of content.[16] Moreover, the implantation of computers into schools, according to its critics, has contributed to an
ideological "nightmare" that promotes certain visions of "what
counts" as knowledge, uncritically accepts that the purpose of
schooling should be to secure future economic competitiveness,
and dehumanizes learners by positioning them as "human capital" or mental components of a "man-machine system."[17]

In short, a concern for "official knowledge" and the "intellectual center" of schooling has been replaced by skills,

competencies, brainpower, and the "science of 21st century learning," and this move has obscured the knowledge-based curriculum from large areas of educational debate. What might this mean in practice?

Soft Openings

In 1993, the British think-tank Demos launched its quarterly magazine with a feature on the future of education, focusing in particular on the work of Howard Gardner, then co-director of Project Zero at the Harvard Graduate School of Education. In his essay entitled "Opening Minds," Gardner articulated the concerns of a "wave of reform" that was dissastisfied with "overblown bureaucracy" and appalled by the uniformity of "school knowledge" with its emphasis on logical and linguistic intelligence. Gardner's recommendations for the design of the ideal school of the future included a more expansive view of multiple intelligences, and a "student-curriculum" brokerage system that would help to match students' profiles, goals, and interests to particular curricula and styles of learning, a task for which interactive technology seemed to offer considerable potential. Many of these ideas were the subject of ongoing development and research at Project Zero.[18]

The ideal vision of a negotiable and flexible curriculum proposed by Gardner in his "Opening Minds" essay were later realized in a major curriculum development program, also called Opening Minds, launched as a pilot project in the United Kingdom in 1999 by the Royal Society of Arts, Manufactures and Commerce (RSA). Here is a concrete example of the globalization of curriculum reform ideas beyond the usual institutional organs and state boundaries. Openings Minds was originally intended to explore a new curriculum model for the twenty-first century, one that put the personal skills, needs, and competencies of learners first while also emphasizing the skills of information handling and knowledge management required in a changing economic and working environment. Opening Minds built upon ideas elaborated in the Gardner essay, aligning them with the RSA's history of intervention in the future of work, "enterprise education," and "education for capability."[19]

Opening Minds is emblematic of a particular type of curriculum reform that emphasizes a "softening" and an "opening up" of the curriculum to both the alleged training needs of the knowledge-based economy and the individual needs and interests of children themselves. Rather than focusing on academic "performance," the specialization of subjects, skills, and procedures and the selection, sequencing, and pacing of pedagogy by teachers, Opening Minds offers a "competence" curriculum. Its competencies approach

refers to a complex combination of knowledge, skills, understanding, values, attitudes and desire which lead to effective, embodied human action . . . at work, in personal relationships or in civil society . . . . Competence implies a sense of agency, action and value . . . The spotlight is on the accomplishment of 'real world tasks' and on a multiplicity of ways of knowing—for example, knowing how to do something; knowing oneself and one's desires, or knowing why something is important, as well as knowing about something.[20]

Competence is realized in the form of projects, themes, and experiences, with learners given greater apparent control over the selection, sequence, and pace of their learning. Competences theories articulate learning as an active and creative practice of constructing personally authentic meanings and understanding and regulating the self; competencies curricula are therapeutic and introspective, empowering and emancipatory. The theoretical and practical origins of competence lie in the 1960s and 1970s, when social scientists and radical educators alike began to celebrate the active, creative, meaning-making potential of individuals—it shares its intellectual origins with the learning sciences—but it is now articulated as behavioral competences and personal learning profiles.[21]

Opening Minds emphasizes five categories of competence: (1) learning how to learn, thinking systematically, creative talents, and handling ICT and understanding its underlying processes; (2) citizenship, ethics, and values, cultural and community diversity, and understanding social implications of technology; (3) relating to people, teamwork, communication, and emotional literacy; (4) managing situations, time management, change management, being entrepreneurial and initiative-taking, and managing risk and uncertainty; and (5) managing information, accessing, evaluating, differentiating, analyzing, synthesizing, and applying information, and reflecting and applying critical judgment. In practice, Opening Minds is usually arranged as a series of thematic and cross-curricular projects.

By early 2012, over two hundred schools officially run some form of Opening Minds competencies curriculum, the program includes its own showcase school and a network of best practice "family schools," and it has been spun off as an independent organization. It has generated a related "area-based curriculum" approach focused on building curricula programs from the needs of specific localities and communities. Instead of being a centrally managed, bureaucratic, and uniformly programmed curriculum, Opening Minds has been interpreted and enacted in multiple different ways. It is promoted as a curriculum framework to be recontextualized according to the specific ethos and history of each school that adopts it, and it positions teachers as creative curriculum actors rather than merely its relays. It has become a well-known "brand" in the UK educational marketplace, with new schools required to pay a subscription fee for participation.

The soft openings of the curriculum embodied by Opening Minds signify a greater porosity and interpenetration between school knowledge, vocational knowledge and skills, and everyday knowledge. Whereas the traditional curriculum associated with conservative restorationism has tended to drive centripetally inward toward a common core of academic knowledge, the soft openings approach develops centrifugally outward into economic and cultural domains. The competencies framework switches together an entrepreneurial vocabulary of initiative, risk, teamwork, brainpower, and so forth with a civic discourse of community values, empowerment, and cultural diversity. Flexibility in the Opening Minds curriculum allows learners to concentrate on interconnected contemporary topics, community sources, and real cultural contexts.

Boundless Creativity

A complementary approach is advocated in the United States by the major Partnership for 21st Century Skills, an advocacy coalition with members from all the major multinational computing, media, and educational services corporations. The mission of P21 is to promote "21st century student outcomes," which it defines through a wide-ranging analysis of learning science theories and their appropriateness for life and work in the global informational and economic landscape. P21 has wide acceptance in the business community, was initially funded in 2002

with $1.5million from the US Department of Education, is con-
nected to many state departments of education through its State
Leadership Initiative, and in 2011 produced bipartisan policy
guidance on "21st Century Readiness for Every Student" that
was introduced in both chambers of Congress.[22]

P21 draws its conceptual and intellectual momentum from
a heterogeneous mixture of sources and associations (though
perhaps its most obvious point of comparison is A Nation at
Risk, with which it shares concern for American global competitiveness in a flat world but which it does not reference at all).
In the white paper setting out the mission and vision for P21,
progressive educator John Dewey is cited approvingly, along
with pioneering psychological work on constructivism-, a range
of cognitive science perspectives, frameworks of creative skills,
emotional intelligences, and multiple intelligences, and assorted
media and technology theories from the 1960s to the present.
These theories are switched together with discourses of "boundless creativity," innovation, and competitiveness in the global
economy. P21 sets out to promote boundlessly creative and
innovative learners.

Accordingly, the necessary skills and "multidimensional"
abilities to be mastered include (1) creativity and innovation, including creative thinking and acting on creative ideas;
(2) critical thinking and problem solving, including the ability to use reason, use systems thinking, and make judgments
and decisions; (3) communication and collaboration, including teamwork; (4) information and media and technology
skills, including information management, media analysis, the
creation of media products, and using ICT for research and
appropriate networking; and (5) life and career skills, especially
flexibility and adaptability, initiative and self-direction, social

and cross-cultural interactions, productivity, and leadership and responsibility. These are all framed by "interdisciplinary 21st century themes" that address global issues, finance, economics, business and entrepreneurship, civics, and personal and environmental responsibility.[23]

P21 acts as a connecting switch between the emancipatory and empowering discourse of constructivism and creativity and the economic discourse of competition that has its origins in the apparent crisis of American schooling to meet the changing needs of industry—here is a nation at risk, once again, in a flat world of global connectivity. It presents a vision of boundary-free creativity, supported by emerging scientific theories of learning, as the panacea to this crisis. The P21 framework is a recipe for a high-tech competencies curriculum.

Despite clear differences with Opening Minds in the United Kingdom, both programs contribute to the same blend of innovation and personal emancipation, as well as a reorientation to knowledge and learning. Knowledge is reconfigured as thematic, modularized, connective, boundary-free, hybrid, and generic; learning is reconfigured as competence, thinking, problem solving, and "learning to learn." This is in line with the style of thought associated with advocates of the knowledge economy: competitive advantage is to be secured by integrating people's behavioral competences with modularized task components.

Curriculum Hybridity

The soft openings trend in curriculum design, as shown by Opening Minds and P21, promotes learning that will prepare students to deal with cultural, economic, and technological change. Disciplinary knowledge and subject expertise has been marginalized by such future-focused agendas. Whereas subject knowledge is organized according to the principle of insularity, its difference from everyday or commonsense knowledge, the soft open curriculum for the future is organized according to principles of connectivity and hybridity. Connectivity and hybridity reject the importance of boundaries between subjects and disciplines, and educational hybridizers instead argue for greater integration and blurring between academic, workplace, and experiential learning. Curriculum connectivity and hybridity celebrates malleable boundaries, integration, and interpenetration.[24]

The soft openings trend represented by Opening Minds and P21 is continuous with international policy agendas that put the emphasis on the brainpower and human capital required by the future knowledge society. International comparative tests and studies of educational performance undertaken by the likes of the OECD and the International Association for the Evaluation of Educational Achievement (IEA) demonstrate how the competences and skills associated with the soft openings trend have become a global testing standard to allow politicians to assess their national performance and achievements against competitors.[25] These comparative instruments are perhaps the "hard openings" to the soft openings of Opening Minds and P21.

The basic assumptions underlying the argument for hybridity have been criticized both theoretically and empirically. From the theoretical perspective, it is argued that a curriculum is impossible without a clear separation of school knowledge and experiential everyday knowledge. Simply put, the idea of a curriculum is to support students' acquisition of new knowledge that they cannot gain through experience. Experience may be a powerful source, but it is no basis for reliable knowledge or a curriculum.[26] Empirical studies have also queried the assumptions of the soft opening trend in curriculum. Such studies show that the high-tech/high-skills/high-wage future promised by such programs has largely turned out to be imaginary. In fact, there is now a worldwide surplus of highly educated graduates—raw brainpower—who are unable to win jobs commensurate with their qualifications.[27]

This chapter has begun to address questions about what knowledge is to be included in the curriculum of the future, what legacies it draws on, and what futures it envisions. The soft openings trend represented by Opening Minds repositions knowledge as "competence" while P21 stresses informational skills. These programs give authority to new ways of knowing and new forms of brainpower that are understood to be more relevant and appropriate to life and work in the digital age, although these assumptions have been questioned on both theoretical and empirical grounds. The next chapter looks for alternative examples of possible curricula of the future, locating them in an "open" world of complex network systems.

3   Networks, Decentered Systems, and Open Educational Futures

Whereas the soft openings style of thinking about the curriculum examined earlier has emerged from a mix of behavioral competence and business innovations, this chapter focuses on an emergent curriculum ideal of networked connections, complex systems, and "open education." It examines how ideas about learning in an emergent open educational commons are linked to questions about the curriculum. Key issues raised by the networked version of the curriculum of the future are those having to do with the connectedness of knowledge areas and how they are defined and with the connections between the curriculum and the kind of society desired for the future.

The Death of the Center

The concept of networks has assumed huge significance as a twenty-first-century style of thought. The language of our times, it has been claimed, talks of systems, complexity, feedback, matrices, lateral connectedness, associations, hybridity, fluidity, multidimensionality, and connectivity. Networks do not only take the form of electronic communications (they are of course a very old form of social organization), though it is in the realm of the high-tech that networks have really entered the public imagination.[1] In comparison to the twentieth-century industrial era of mass production, centralization, and organized hierarchy, pinpointed by the image of a single central dot to which all strands led, the twenty-first century digital age has been defined by the "death of the center" and its replacement by a mesh of many points all linked multidirectionally to webs and networks. The current era is characterized by the plasticity of information, the perpetual beta, an open, decentralized approach to information, and open-source politics, all powered by the Internet's centrifugal forces.[2] In such a smart decentralized world of networks, it is argued that the dynamic and the mobile are challenging centralized bureaucracy, dialogue and cooperation are preferred to hierarchical authority and order, flexibility seems more important than routine, and a counter-culture of the Internet geek has taken over for the dark-suited manager of the big firm. Twenty-first century society is a lateral society of fluid networks rather than a vertical society of totalizing structures.[3]

Network-based technologies introduce new possibilities for interaction, common dynamics, and participation into everyday life and learning. As a result, researchers working in the field of digital media and learning have explored the significance of "networked publics." Networked publics refer to the intersections of domestic life, nation-state, mass-culture and commercial media, and everyday life in the context of a convergence of mass media with online communication. Networked publics, like many other types of publics, allow people to gather for social, cultural, and civic purposes, and they help people connect with a world beyond their close friends and families. As a result, networked publics now increasingly constitute the social groups that structure young people's learning and identity. They provide opportunities for engagement in hobby-based or "interest-driven" publics that exist outside school or existing friendship networks.[4]

According to research on networked publics, learning is now increasingly decentered and dispersed in time and space, horizontally structured, networked and connective, and convergent across many different media. In a networked world, learning can take place online as well as in high schools, museums, after school programs, homes, business, broadcast media, public libraries, and community settings. The emphasis is increasingly on dispersed, decentralized, and virtual learning taking place fluidly across lifetimes, social sectors, and media, with the Internet itself imagined as a learning institution. Such arguments are set against schools understood as innately conservative institutions that continue to rely on structured hierarchical relationships, a static print culture, and old-style transmission and broadcast pedagogies that are at odds with the networked era of interactivity and hypertextuality.[5]

At the same time, developments in the "open access" of information and knowledge in higher education and scholarship have begun to point to radical new possibilities for schooling. Open access means putting peer-reviewed scholarly material on the Internet to be made available free of charge and free of most legal restrictions. Some major research universities have pioneered open access as a way of bringing down the public barriers to research. MIT led the way with OpenCourseWare, while Harvard University's faculty of arts and science has adopted an open archiving mandate.

In the emerging "open education" paradigm, educational materials are digitized and offered freely and openly to educators and learners to use, customize, improve, and even redistribute in their own teaching, learning, and research. A series of major reports has advocated for open education in the United States and Europe, contributing to the establishment of new "knowledge ecologies," "knowledge cultures," and a "global knowledge commons" based on a new collection of values of openness, an ethic of participation, and an emphasis on peer-to-peer collaboration. Open education is an educational paradigm for a seemingly "open era" based not only on a technological discourse (open-source, open systems, open standards, open archives, and so forth) but on a change of philosophy that emphasizes ideals of freedom, civil society, and the public sphere.[6]

Consequently, arguments in favor of informal networked learning and arguments for open education have been enrolled into arguments advocating for curricular change. The following case studies exemplify the potential for openness in the connected curriculum of the future.

Systems Curriculum

Quest to Learn in New York City offers a blueprint for a possible future of institutional schooling after the death of the center. The school's main documents emphasize "systems thinking" and "learning about the world as a set of interconnected systems," and it is "committed to graduating strong, engaged, literate citizens of a globally networked world." Based on this strong systems language, it reimagines "school as just one kind of learning space within a network of learning spaces that spans in school, out of school, local and global, physical and digital, teacher led and peer driven, individual and collaborative."[7] Quest to Learn (Q2L) is an ideal-type school for a dispersed field of interest-driven learning in networked publics.

The entire Q2L experience is designed around the notion of "game design and systems." It establishes the architecture and culture of videogames as its core principles for curriculum design. This does not mean that the student experience involves a lot of playing videogames. Instead, the learning experience is designed according to the principles of videogame design. In turn, it assumes that videogame design embeds effective learning principles in highly motivating contexts. Q2L is an institutionalized version of the argument that good videogames make effective learning machines. For example, videogames present players with problems to solve that are designed to become progressively tougher to solve, offer continual feedback on progress, are customizable according to different styles of play, enforce repeated cycles of practicing skills as a strategy for accomplishing goals in authentic contexts, and offer intriguing situations and characters that require deep affective player investment.[8]

Moreover, according to Q2L documentation, videogames constitute an ideal technology for promoting systems thinking. Systems thinking refers to the understanding that any system—social, technological, natural—maintains its existence and functions through the dynamic interaction and interdependence of its parts. Systems thinking stresses the unintended consequences of complex interactions and relationships. It is antithetical to the traditional curriculum of insulated subjects, isolated facts, and knowledge learned out of context. As complex systems, videogames are positioned at Q2L as a more appropriate medium for the future of learning than a conventional curriculum of separated subjects and linear knowledge domains.

In order to support its systems thinking focus, Q2L "posits learning as context-based processes mediated by social experiences and technological tools," a "highly social endeavor" that takes place through "situated practices" within "communities of practice":

In this way, a situated-learning view stipulates that learning cannot be computed solely in the head but rather is realized as a result of the interactivity of a dynamic system. These systems construct paradigms in which meaning is produced as a result of humans' social nature and their relationships with the material world of symbols, culture, and historical elements. The structures, then, that define situated learning and inquiry are concerned with the interactivity of these elements, not with systems in the individual mind.[9]

Through this approach, students at Q2L are engaged in situated and authentic, real-world learning experiences. The distinct Q2L conceptual framework for the curriculum hybridizes the systems language of videogames design with the systems language of situated cognition derived from the learning sciences.

Besides the systems focus, Q2L also has a strong emancipatory ethos. It positions its students as "sociotechnical engineers" who can create systems (games, models, simulations, stories). By "designing play," it claims, "students learn to think analytically, and holistically, to experiment and test out theories, and to consider other people as part of the systems they create and inhabit." The built-in creativity and design focus seeks to produce students who are empowered to act and make and participate in global dynamics rather than receive and consume.

In order to do so, Q2L also provides a coherent, structured curriculum model that claims to juxtapose state standards with twenty-first century skills. The curriculum is organized as interdisciplinary knowledge domains instead of separate subjects.

Each interdisciplinary domain structures learners' experience in integrated expertise such as researching, theorizing about, demonstrating, and revising new knowledge about the world and its constitutive systems.

The integrated domains are described as follows. "The way things work" integrates science and math and involves learners dismantling different kinds of systems and modifying, remixing, and inventing their own. In the "being, space and place" domain, students study the social, temporal, and spatial forces that shape the development of ideas, expression and values through combinations of social sciences and English language arts. "Codeworlds" blends language arts and math and computer programming and involves students decoding, authoring, and manipulating meanings through the symbolic codes, including those of literacy, numeracy, and computation. "Wellness" situates personal, social, emotional, and physical health within systems of peer groups, family, community, and society. Finally, "sports for the mind" emphasizes the fluent use of new media across networks for careers and civic engagement in the twenty-first century. The interdisciplinary curriculum is delivered through problem-based "missions," "levels," and "quests" that are organized according to basic videogame architecture.

Q2L's integrated curriculum embodies a form of networked, collaborative, digital interdisciplinarity. Its keywords are "systems," "dynamics," "integration," and "hybridization," and it seeks to prepare students for a world it characterizes as globally connected and complex. To act in such a world, students need to be able to recognize patterns and identify structures, think connectively and creatively, be inventive and innovative, adopt and tolerate multiple cultural perspectives, exhibit empathy and reciprocation, understand what it means to be an active global citizen, understand and respect the self and others, and understand the various modes of new media communication.

Despite the high-tech, digital interdisciplinarity discourse of game design, then, it is also constituted by a more affective, emotional, and ethical discourse. Q2L is a smart, open, dynamic curriculum of the future that nonetheless continues to resonate with a much longer curricular legacy in the United States. The basic intellectual architecture is derived from John Dewey's insistence on "inquiry," "experience," and "learning community," as remixed through the discourse of open systems and networks and an emphasis on sociotechnical engineering. It amalgamates participation in the economic sphere with notions of community and local responsibilities in the cultural sphere. The first is promoted through emphasizing technological competence and the soft skills required for flexible working; the latter through appealing to authentic and learner-centered or "personalized" learning. It offers a hybrid language of learning that is both high-tech but also emotionally "high-touch."[10]

Additionally, Q2L's curriculum for the future represents the world in terms of complex open systems. Q2L's version of complexity theory emphasizes emergence, nonlinear dynamics, uncertainty, feedback loops, self-organization, and interconnection. In complexity terms, learning, curriculum, and knowledge are understood as continuous invention and exploration, produced through complex interactions among people, action and interaction, and objects and structural dynamics that all produce emergent new possibilities. New problems emerge at the point of solving others; knowing emerges with the appearance of new problems as we participate in the world.

Put educationally, complexity theory promotes a sense of openness and permits the possibility of alternative futures. A complexity curriculum is open, dynamic, relational, creative, and systems-oriented, it involves processes of cross-fertilization, pollination, and the catalyzing of ideas to form a webbed network of connections and interconnections, and it emphasizes learning not through direct transmission from expert to novice, or from teacher to student, but in a nonlinear manner through a class exploring a situation/problem/issue together from multiple perspectives.[11]

The complexity approach taken up in Q2L treats curriculum not as product for imposition but as a process of emergence and interaction. It is forward-looking in that it embraces the contingency and uncertainty of educational outcomes. It recognizes processes of inquiry and exploration, and it mobilizes a vocabulary of networked interactions and webbed learning. The curriculum, from a complexity perspective, is an open system of constant flux and complex interactions rather than a closed system of prescriptions and linear progressions. A complexity curriculum emphasizes students as knowledge producers, organizing and constructing knowledge as they interact, an argument that resonates surprisingly with the political "pedagogies of the oppressed" of Paolo Freire and the radical progressivism associated with John Dewey.[12]

Networked Neoprogressivism

The complexity curriculum of Q2L remixes an emancipatory politics of participation through the globally dynamic, complex systems of networked society. Yet it is certainly not alone in mixing up technical and progressivist codes for thinking about the future of learning. It is part of a vision that might be termed "networked neoprogressivism." Networked neoprogressivism consists of a set of statements and practices that articulate the future of learning in terms of self-organizing webs of activity blended with a reinterpretation of progressivist educational values and aspirations.

For example, the New Basics curriculum program trialed in Queensland, Australia, was explicit about its theoretical roots in radical progressive pedagogy. A booklet for teachers draws from radical progressivist theory, alongside sociocultural psychology, to craft an approach that requires the solution of "substantive, real problems" in learners' worlds, includes "integrated, community-based tasks," and involves teachers as "mentors" scaffolding" the activities of "novice" students.[13]

Elsewhere in the project documentation, the New Basics is conceived in dynamic networked terms. Rejecting the curriculum as a "central authority" based on "economies of scale for publishing, distribution and implementation of texts using print media," the project advocates for "using online, interactive technology for local, regional and global curriculum development and renewal" and the "rapid prototyping, development and revision" of more specialized materials based on "economies of scope." Again, a dynamic decentralization discourse associated with the Web is synthesized by the New Basics documents with a progressive, emancipatory vocabulary of real-world problem-solving.

Perhaps the most radical neoprogressivist view is of a future "post-school era" where the formal institutions, personnel, instruments, and resources of education have been replaced by self-disciplined learning collectives, crowds, and communities, all connected by the Internet. In an imagined post-school era, schools disappear as young people increasingly learn through networks, drawing on personal and domestic digital technologies as sources of knowledge and ways of connecting with others. Instead of prizing disciplinary knowledge, a "curriculum 2.0" acknowledges experiences such as collaborative learning, personal development, self-monitoring, creativity, and thinking skills.[14]

The ideal of an "open source curriculum" put forth in the curriculum 2.0 vision values teachers and learners participating in a wiki culture of production and collaboration over learning materials and resources. The neoprogressive, connectivist curriculum 2.0 is rooted in a pervasive digital discourse of 24/7 learning, nomadic learning networks, transmedia convergence, smart mobs, crowdsourcing, user-generated content, opensource, DIY media, cloud culture, and so forth.

The School of Everything, for example, is a simple Web platform that allows anyone who has something they can teach to link up with anyone who would like to learn it. Its founders describe it as a response to the outdated rigidity of school, and they cite the key source of inspiration as the Free U in 1960s California. It aspires to promote a culture of informal teaching and learning. The School of Everything "manifesto" mixes an empowering people-centered appeal—the concept that "everyone has something to teach," "everyone has their own way of learning," "all subjects are important," and that "people are brilliant, inspiring, generous, and smart—with a critique of "expensive" formal education and of "overrated" qualifications and credentialism.[15]

The post-schooling scenario reanimates the countercultural "deschooling" agenda of the 1970s for the era of eBay and MySpace, reaffirming its attack on institutionalized schooling, its assault on assembly-line learning, and its commitment to self-determined learning through informal networks and community bonds. The radical idea of learning webs imagined by deschoolers is now, it seems, more realistic as learning networks are made possible through the Internet to society as a whole. A much more convivial new hidden curriculum, like the deschooled society of progressivist imagining, facilitates communication, cooperation, caring, and sharing between free agents and distributes learning into a nomadic network of authentic practices, cultural locations, and online spaces.[16]

According to these views, isolated and insulated educational institutions are now being challenged by a much more pedagogically polygamous range of incidental, non-institutionalized learning relationships and attachments. The result has been a restructuring of the spatial and temporal boundaries of education, with learning to be extended beyond learning institutions into virtual environments and stretched across the life span instead of concentrated in youth. All boundaries between informal, interest-driven and formal education are imagined as increasingly flexible and even porous. Formal learning is imagined to be optional or flexible in terms of attendance. Learners are imagined as taking more control over the selection of learning resources and sources, with learning content more customized, malleable, and adaptable. New spatialities and temporalities of learning are opened up by the flexing of timetables, the compression of space by real-time digital communication, and the virtual erasure of school walls. Schools are reconceived as learning spaces designed to afford different ways of working (team working, personal reflection, information access) rather than organized rigidly around faculties and subject disciplines. Learning is decentered and reimagined to be taking place fluidly and flexibly in a utopian dispersed network of formal settings and informal media environments. Networked neoprogressivism is a connective utopia where anything goes with anything else!

Whether the desire for a "technical fix" expressed in a post-school utopia will, however, lead to the high-tech deschooling of society, "leaving us all enmeshed in Illichian webs and nets," is debatable, and it seems more likely that education will continue to be "framed within the competing claims and complexities of democracy and capitalism."[17] The idealization of networks in the imagining of the connective curriculum of the future, therefore, needs to be understood critically. The connected curriculum of the future is no value-neutral or depoliticized utopia: it is enmeshed in complex social, economic, and cultural trends. For starters, network discourse is a form of technological determinism that reduces all other phenomena, relations, and forces to the logic of technological change.[18]

More particularly, critics claim network technologies have brought about a greater emphasis on fast time and short-termism over long-term thinking, while the reality for many teachers and learners remains that of centralized and hierarchical "techno-bureaucracy" rather than open educational "cyberpedagogy." Decentralized control over curriculum and learning resources is not always liberating, but may bring about disunity, disconnection, desolidarization and disadjustment, dysfunctionality, destructive conflicts, exploitation, and other negative effects. The network-centric and horizontal utopia of the future of education systems tends to flatten out and glide over persistent educational inequalities and asymmetries; it idealizes community, respect, equal power, and entrepreneurialism, but it elides over disciplinary problems and differences, reduces knowledge to marketable commodities in the form of "soundbites," and reimagines education as "learning bubbles."[19]

Wiki-fied Futures?

Informed by network thinking, centrifugal schooling lashes together and hybridizes a range of "open" educational theories and ideas about complex networked systems into an emergent way of thinking about schools in the twenty-first century. An emergent, open, networked ideal of curriculum design for centrifugal schooling is now part of a twenty-first-century style of thought about the curriculum that consists of concepts such as complexity, connectivity, convergence, emergence, interactivity, openness, playfulness, systems, and webs. This style of thought does not only seek to explain a new social world to which the curriculum ought to be reformed; it helps to construct that world, as seen at Q2L. Q2L uses cutting-edge pedagogical and instructional techniques, twinned with innovative technologies, to create new kinds of learners with new ways of thinking, seeing, and practicing in the world.

The generalization and idealization of the learning benefits of networks into a style of thought is an aspect of wider social, economic, and cultural changes. Its emergence is shaped by the interactive effects of globalization and the digital revolution as well as by economic restructuring processes that drive privatization, deregulation, and open markets.[20] The mindset of computer engineers and the entrepreneurial hacker culture of Silicon Valley—the cyberlibertarian "California ideology" as it's sometimes known—has now diffused throughout popular culture and worked its way into the styles of thought, the minds, and the imaginations of the public, as well as into the business plans of transnational media companies.[21] The smart networked vision of the curriculum draws its impetus from "apologists for the flattening of the world, and bureaucratic enforcers of the proclaimed new global order" who envisage society in terms of benevolent network connections and relations.[22] They channel a new vocabulary of "wikinomics" and "wikicapital" associated with deregulated "open markets" into the new soft logic of learning.

For example, major supranational organizations like the OECD and UNESCO, as well as transnational computing corporations, all now spearhead programs encouraging open education based explicitly on the interactive culture of the Internet and the utopian ideal of user-generated knowledge embedded in YouTube and Facebook to produce a kind of "democratic" wikified vision for the future of the curriculum.

Yet letting the "geeks of Silicon Valley" make decisions about education, albeit at a distance, may mean that the "future of education in the digital age will be determined by our judgment of which aspects of the information we pass between generations can be represented in computers."[23] The implications for curriculum are significant. If the curriculum is a relay of knowledge between generations, then a reduction of this relay to only media that can be computerized has the potential to exclude significant cultural materials and to promote narrowly specified ways of being and thinking.

New technologies have therefore been criticized as part of a "politics of public miseducation" in the curriculum, "the latest technological fantasy of educational utopia, a fantasy of 'teacher-proof' curriculum" that eschews "interdisciplinary intellectuality, erudition, and self-reflexivity."[24] According to such critiques, network discourse and rationality has begun to install in education particular kinds of design decisions and algorithmic assumptions that are rooted in the logic and embedded values of computer engineering rather than in the intellectual

concerns of educators. The emerging style of curriculum thinking is a wiki-fied geek style originating from well outside the
normal institutions and mindsets of educational systems. This
points to the need to understand how new actors from outside
the usual institutions of the education system—and the politics and values they catalyze—are now involved in educational
designs for the future, as the next chapter shows.

4 Creative Schooling and the Crossover Future of the Economy

Earlier it was shown how the ideas underpinning some examples of the curriculum of the future are continuous with the ideology of securing future competitiveness in the knowledge economy. These ambitions have been reinforced in the wake of the global recession. This chapter goes on to explore more specifically how the curriculum of the future is being imagined and constructed through the work of the private sector actually working inside of public education. The future of the economy and the curriculum of the future are now being reassembled together through public-private partnerships.

The argument is that the curriculum of the future and the economy are networked together through all kinds of mergers of public and private and state and commercial sectors and objectives. In the twenty-first century, education systems are not controlled by a centralized government authority but through a decentered network of various authorities. A new kind of "polycentric" or "multipolar" educational politics is emerging in which education is done through hybrid mixes of public and private bodies, bureaucracies, and markets rather than by one single center of authority.[1] Again, these are aspects of a new style of thought now used to explain the problems of the curriculum and to intervene in its future. This chapter asks how curriculum policymaking gets done in a polycentric context, and how this might affect the makeup of the curriculum of the future. What specific networks of organizations and individuals, and cross-sectoral and interorganizational connections, are involved in imagining the future of the curriculum, and to what purposes and ends?

Schooling to Work?

Sociologists of education in the United States and the United Kingdom have for many decades debated the links between the school curriculum and employment. Classic studies of the 1970s posited clear correspondences between the social authority of the curriculum, the socializing and sorting function of schooling, and paid work in the capitalist economy. Advocates of the theory of "human capital" argued that education should be thought of as "investing" in human resources that would later benefit the national economy, and therefore that curricular content should focus on preparation for employment and the needs of industry.[2]

In the context the knowledge economy described earlier these correspondences are harder to detect, but human capital theory remains a powerful political influence. In the knowledge economy, workers are required to be creative and "flexible specialists" who can adapt to fluctuations and changes in market demands. This makes quite a few new social, intellectual, and educational demands of employees and thus of schools. To reiterate points made earlier, both schools and businesses now speak the same language of flexibility, modularization, componentization, competences profiles, soft performance, brainpower, problem solving, and so on.

For instance, the High Tech High charter schools network established in Southern California in 2000 was conceived by civic and high-tech industry leaders in San Diego, and assembled by the Economic Development Corporation and the Business Roundtable, to discuss the challenges of preparing individuals for the high-tech workforce. Its aims include the integration of technical and academic education to prepare students for participation in high-tech fields. HTH describes itself as an "open-source" organization that offers free resources and services for other educators. It "places a premium on retaining flexibility and agility," and it emphasizes the importance of its "collective undertakings," caring culture, and the preservation of the organization's "soul."[3]

It is clear that High Tech High, like other prototypical examples of the curriculum of the future, is concerned with students' future employment and it adopts the flexible correspondence model that flexible learning = flexible labor. Particularly in light of the global recession, however, some of these arguments have been softened and programs like HTH have adopted a more "soulful" language of creativity. Both in its objectives and in its textual presentation, HTH mobilizes a high-tech language enriched by a more humanist organizational soulfulness. Likewise, Opening Minds, Learning Futures, and Quest to Learn are all compelling examples of new curricula that promote capacities for innovation required in a new economy through a discourse of "reenchantment."

According to this reenchantment, the new economy of the twenty-first century is more socially responsive, ethical,

compassionate, customer-facing, fun, and informal. It is characterized by its nonconformist countercultural "cool" and a seemingly anticorporate "hacker" spirit of rebellion and individual
liberty. Most of all it represents an "age of creativity" in which
being creative is considered the highest achievable good. Relent-
less innovation, and 24/7 productivity are now the chief characteristics of the creative types who inhabit this age of creativity.
The latest technological gadgets are enrolled in this anticorpo-
rate-capitalist universe of cozy techno-bohemian work-life bal-
ance. Corporate capitalism is no longer to be associated with the
9–5 businessman in the dark suit but with the restless creative
entrepreneur dressed in black.[4]

In this creative universe "affective labor" takes place "in-
person," engendering "feelings" such as ease, well-being, sat-
isfaction, excitement, passion and so forth, and distinctions
between leisure, labor, domesticity, sociability, production and
consumption become blurry.[5] Affective labor and creativity in
the digital economy displace faceless bureaucracies with a caring and sharing capitalism, or business with personality. In
other words, we have now been "taught that corporations have
a soul."[6] In this "creativity explosion" business culture values
creativity over routine, and education seeks to promote in children the creativity required for nonlinear thinking and generat-
ing new ideas.[7]

The creative and affective reenchantments of the economic
domain are mirrored, then, in the educational domain. Projects including Opening Minds, Learning Futures, and High Tech
High are evidence of how the economic emphasis on effectiveness, efficiency, accountability, measuring, and so forth has been
softened by a more cultural focus on empowering learners, elic-
iting learner voice, and paying attention to learners' emotions.

Keywords of the reenchanting vocabulary of schooling are "happiness," "well-being," "emotions," and "self-fulfillment." Effectiveness is replaced by "affectiveness" and by the "expressiveness" of creativity. The change is something like a shift from an "asset management" language of "bastard leadership" in schooling where students are treated as assets to the school—a form of human capital to be virtually exchanged for competitive performance table positions—to a new language of "affect management" and caring leadership. In the affect management style, schools are responsible for the monitoring, regulation, and control of students' emotional selves. The aim of schooling is to produce well-adjusted emotional selves who can take ownership, feel empowered, be creative, and experience enjoyment of learning. This requires affective schools rather than effective schools, and the production of passionate, feeling, affective learners.[8]

Learning to Playbor

Consequently, another aspect of the reenchantment of educational language has been its appeal to young people's existing digital cultures and their informal learning with new technology—lessons already learned by the leading "cool" companies of the new soulful economy. The successful "leading-edge 'techy' organisations" are already "tapping into the skills developed by a generation that has grown up with Nintendos, Xboxes, and more recently online multiplayer games."[9]

For example, videogame companies have successfully recognized that the "work ethic" of routine, restraint, stratification, and deferred gratification can be replaced by a "play ethic" of "passions" and "enthusiasms" and "feelings." There has been a thorough hybridization of the "playground" and the "factory" in Internet culture and the interactive economy. The merging of play and work has resulted in "playbor," a neologism that accurately captures the ways in which the affective elements of play have now been merged into the value-making tasks of the economy.[10]

Whereas the old model of schooling to work involved learning to labor, the curriculum of the future is concerned with learning to playbor. To illustrate, Quest to Learn focuses on playful systems and the important role of videogames in introducing players to the complex skills required in the twenty-first century. Enquiring Minds and Learning Futures both work with a "learner voice" agenda that gives young people greater autonomy and ownership of their learning. A booklet produced by the Learning Futures program in collaboration with High Tech High speaks of learning being "passion-led," "fun," "exciting," "inspiring"—it should have "real-world" relevance, stretch students' "intellectual muscles" as "expert learners," and "ignite students' imaginations."[11]

The expert learners positioned by these texts are creative playborers whose affectiveness, well-being, and creativity are understood to be essential prerequisites for economic reinvigoration. It is through this reenchanting explosion of creativity that commercial organizations have sought to expand their operations in education, not simply through traditional tactics such as marketing but by working inside public education itself.

Commercialism in the Curriculum

Commercial organizations routinely supply products as diverse as vending machines and textbooks to schools. But this form of commercial activity in schooling and the curriculum is just a small part of private-sector participation in schools. Commercial activities include sponsorship of programs, sponsoring materials, promotion and marketing of software and technology infrastructure, exclusive agreements such as those made with textbook publishers, electronic marketing, incentive programs such as store vouchers, school facilities reconstruction programs, plus the full privatization and management of schools.[12]

Consequently, commercial activities may now shape the structure of the school day, influence the content of the school curriculum, and determine whether children have access to a variety of technologies. Commercialism represents an array of alignments between commercial organizations and education, or the entanglement of politics, education, and private finance in a new world of global for-profit education and knowledge industries. Public education, then, is big business and many critics find this alarming. "Edu-business" and privatization bring the normative assumptions of global market competitiveness into public education, arguably leading to a narrowing of what is seen to count as students' learning. Some schools are even run like companies competing against one another in free markets. These developments are important considerations for anyone involved in understanding the design of the curriculum of the future.

Many major transnational corporations are involved in multiple sites of curriculum innovation. For instance, Futurelab's Enquiring Minds program was funded for four years by Microsoft's "philanthropic" arm Partners in Learning. Microsoft promotes educational innovation and sponsors many specific "innovative schools" worldwide. The Gates Foundation was a key partner in founding the original High Tech High charter school and supports New Visions for Public Schools, one of the principal partners in Quest to Learn. In addition, Microsoft is a member of the Partnership for 21st Century Skills, along with many other companies including Cisco, which also promotes its own vision of "connected schools." Cisco commissioned and published a report on worldwide education innovation by researchers from the same think tank, the Innovation Unit, which also established the Whole Education network with which Futurelab's Enquiring Minds project, Learning Futures, and Opening Minds are all affiliated. By tracing these links it seems that the future of the curriculum is a mobile vision that moves across commercial and noncommercial sites, crisscrossing national and sectoral borders and circulating among multiple agencies and organizations.

These connections are concrete examples of the reach into public education attained by commercial and private-sector companies in recent years, much of it accomplished through philanthropic and corporate responsibility programs and encouraged by the creativity explosion. While corporate philanthropy in education clearly has its merits, critics remain concerned that the building of public goodwill and positive brand image through corporate responsibility constitutes covert advertising and marketing in schools.[13]

Crossover Governance

Commercialism in the curriculum of the future is one specific aspect of a less visible and more complex phenomena in public education captured in the term "soft governance." Simply put, there has been a shift from the "hard government" of legislation, regulation, monitoring and compliance to the "soft governance" approach of recommendations, education campaigns and strong advocacy (although the extent of the transformation is debatable), leading to a blurring of the distinctions between the public institutions of government and the work of private companies. Public policy making has expanded to include individual actors, companies, social groups, civic organizations and policy makers that all interact with each other in a multilayered, multidimensional and multi-actor system to bring about collective goals. At the same time, soft governance has been accompanied by a shift from hierarchy to networks. The new networked or cellular relationships involved in education involve both public and private sector players as well as those located in between, especially non-profits, charities and other "crossover" organizations that crisscross, straddle, and bridge sectoral boundaries.[14]

The emphasis on networks in governance describes processes that are decentralized and characterized by fluidity in order to cope with rapid social change, intense societal complexity, and instability. Working through networked structures, the political center of government encourages cross-sectoral participation with nongovernment actors but retains a coordination or steering function over policy. The state tenders for contracts, outsources services, and monitors delivery, but does not necessarily manage education services directly. Within education policy, it is argued, networks serve as a way of trying out new ideas, getting things done quickly, and interjecting practical innovations and new sensibilities into education.[15]

The kinds of networks that now cooperate to get educational innovations implemented consist of nongovernmental and intergovernmental organizations, think tanks, nonprofits and social enterprises, as well as global organizations such as the World Bank, the World Trade Organization (WTO), the International Monetary Fund (IMF), the Organisation for Economic

Co-operation and Development, UNESCO, and the United Nations, and multinational private-sector corporations and their philanthropic initiatives. All these agencies draw their language of expertise increasingly from the logic of networking and open-source organizational models from the Web.

The emergence of soft networked governance has allowed new hybridizations of public, private, and crossover organizations and actors, connected via a range of cross-sectoral and interorganizational networks, to design and deploy a range of novel programs for the future of schools. Here it is possible to detect resonances of the reenchanting discourse of creativity, well-being, and personal affect that characterize companies in the interactive digital economy. Key crossover actors, agencies, and organizations have turned keywords like "creativity" and "innovation" into policy reform slogans and incantations. In the United Kingdom, think tanks like Demos and the Innovation Unit have been early adopters of such slogans. The future of education is to be recast in terms of learners' individual passions, their well-being, and the purposeful creativity of youthful digital pioneers. The kind of innovations required is to be found in open-source hacker communities and in the rapid R&D culture of Silicon Valley.[16]

The curriculum experiments of centrifugal schooling are paradigmatic of soft governance through policy networks of public, private, and intermediary crossover actors and agencies. The Enquiring Minds curriculum project from Futurelab was the product of cross-sectoral networks and soft governance, brought together by a discursive emphasis on affective and creative learning to produce a "crossover curriculum."

Enquiring Minds was intended as a study in curriculum change, where the curriculum is understood as the outcome of interconnections between institutional structures, everyday practices, and policies rather than as a product to be implemented. The program was premised on the idea that teachers and students might be involved in decisions about the content and structure of aspects of the curriculum, and that curriculum making is an ongoing and complex process of constant assembling, dissembling, and reassembling of sources, texts, plans, and schemes of work. It is illustrative of how the curriculum is constituted, assembled, and materialized through a composite of discourses, texts, actors, organizations, interpretations, and diverse materials.[17]

The interorganizational network that produced EM is significant. The project was initiated and run by Futurelab, a not-for-profit educational R&D "lab" intended to support innovation in educational technology.[18] Futurelab was originally established in 2001 by the National Endowment for Science Technology and the Arts (NESTA) with funding from the government Department for Education and Skills (DfES) and later from the quango Becta (the agency with responsibility for information and communication technologies in schools). The vocabulary of Futurelab emphasized innovation, ideas, incubation, collaboration, user-centeredness, personalization, as well as a more technical vocabulary of open source, social software, and, of course, networks. The language of creativity was deployed to bind these entrepreneurial, technical, and learning elements together. Futurelab acted as an intermediary among government, industry, and academia. It deployed methods of educational and technological entrepreneurship and epitomized the social economy and social enterprise ideals of the nonprofit sector.

As already noted, EM was firmly connected to private-sector participation in education. EM was funded by the global "philanthropic" fund of Microsoft Partners in Learning (MS PiL) that aims to "help educators and school leaders connect, collaborate, create, and share so that students can realise their potential." In the United Kingdom, MS PiL has partnered with the devolved governments of England, Wales, Scotland, and Northern Ireland, as well as with Futurelab. Microsoft is also affiliated to the Partnership for 21st Century Skills. Clearly Microsoft embodies commercial as well as philanthropic objectives in its interactions with schools and its sponsorship of a variety of programs trying to influence the future of education. Sponsorship of EM included a great deal of branding, including a glossy printed curriculum guide, a bespoke Web site, and an online tool for promoting inquiry. Not only did funding from MS PiL permit the research to proceed, it also permitted the Microsoft logo to be associated with a new curricular innovation working directly in schools: it positioned the brand within the curriculum.[19]

In terms of its public-sector connections, EM was supported by an advisory group consisting of individuals from the government-funded Qualifications and Curriculum Agency (QCA) and from the RSA program Opening Minds, academics concerned with researching ICT in education, the local government, and an advisor from a leading think tank and the Prime Minister's Strategy Unit. Further meetings were held with staff from a range of other government agencies, commercial companies, and crossover organizations. This list of project advisors and other specialists is indicative of the relations across public, private, and cross-sectoral spaces that now contribute to curriculum design.

EM was also enacted as practice in schools. The original project was concentrated in two specific secondary schools in the southwest of England, but after dissemination of the curriculum guide to over three thousand schools the researchers provided training to teachers in about one hundred schools. It went through a variety of localized or vernacular interpretations. Throughout, the project was characterized by tense exchanges and disagreements, contests and compromises, both between the researchers and the participating teachers, as well as between the researchers and its advisors and sponsors. The project was a site of continual negotiations and attempts to enroll other actors to make links between things and people, and to find consensus in order to acquire some long-term stability and durability.

In addition, EM was part of a UK movement in curriculum innovation that gave rise to Whole Education, an "open-source alliance" of projects dedicated to exploring the future of education through cross-sectoral partnerships and connections. Whole Education, and the projects, programs, and organizations it represents, forms a loose network within which the authority for curriculum planning has been taken up by a range of new sources of expertise that are not associated with the traditional organs of the education system. Instead, Whole Education consists of nonprofits, voluntary and charitable organizations, social enterprises, and think tanks, each supported by a plethora of public and private sources of funding, expert and business advisory groups, philanthropic sponsors, and so forth.[20]

Quest to Learn in New York City, too, is the outcome of a decentralized network of interorganizational and cross-sectoral relations. Its "development process has included a range of partners who bring innovation and credibility to the work."[21] Its partners are positioned as expert participants rather than as sponsors. Q2L was commissioned by New Visions for Public Schools, which proposes that "educational improvement requires everyone involved—the public school system, government, businesses, community groups, parents and students—to work harder and do better together."[22] Since 1989 New Visions has created over 130 new "small schools" and mobilized community groups, institutions, and businesses to support them; it has initiated a school creation program and improvement strategy that has been adopted more widely by the New York City Department of Education; launched a teacher recruitment, preparation, and retention program; and pioneered a principal mentoring program for over six hundred principals in their first year of service. New Visions became a New York City Department of Education "Partnership Support Organization" in 2007 and has begun opening its own charter high schools, thus positioning itself both within and beyond the formal education system. Some New Visions reforms have been replicated in other sites as it expands through the education system. Its funding comes from a mix of government, corporations, foundations, and individual sources. Rather than being seen as an isolated outpost of innovation, then, Q2L's partnership with New Visions locates it in a matrix of interorganizational and cross-sectoral relationships and reforms.

In addition, the conceptual and organizational model of Q2L was designed by the Institute of Play, a games and learning nonprofit staffed by professional game designers and researchers in the field of game-based pedagogy, new media literacy, and the learning sciences. The Institute of Play has been funded by the MacArthur Foundation. Additional collaboration on Q2L has come from Parsons The New School for Design, particularly its mixed-reality lab, and consultation with curriculum and teaching experts, middle-school students, and selected academic experts involved in researching digital media and learning. The creation of Q2L is therefore the result of a network of participants and resources from government, business, community, philanthropy, and academia. Its curriculum has been assembled from a heterogeneous network of elements working upon one another; it brings together, fuses, and freezes in one form a whole variety of voices, explorations, ideas, visions, concerns, conflicts, and alternative possibilities.[23]

These examples exemplify how curriculum reform is now increasingly done through the "good ideas" of "policy networks," consisting of nonprofits, think tanks, quangos, and social enterprises, plus key individual intermediaries, interlockers, and policy entrepreneurs, which straddle the public and private sectors. These networks build consensus about what works in education reform through explicit partnering and dissemination activities, the production of texts, in-house publishing, project Web sites, and online networks. They epitomize "entrepreneurial governance" through "ephemeral networks," partnerships, outsourcing and contracting-out, marketization, and devolution or decentralization.[24] Futurelab and the Institute of Play are good examples of organizations involved in such ephemeral networks, and EM and Q2L illustrate how curriculum innovations can be produced in a networked policy environment.

What are the implications of cross-sectoral and interorganizational governance in the design of the curriculum of the future? Pragmatically, private- sector support and philanthropic sponsorship are at least a financial requirement for trying out new curriculum ideas. In addition, official government partnership or sanction helps to embed these programs within public education sites and spaces.

In terms of curricular content, crossover governance also endorses and produces new kinds of "official" knowledge. Q2L embeds in its curriculum forms of knowledge imported from computer science and video gaming, and it structures the way in which learning takes place according to theories of situated practice, inquiry-based teaching, and complex problem solving that have been articulated in the learning sciences. Its academic domains are crossover domains too, with a great deal of interdisciplinary inquiry a major feature of the curriculum structure. There are strong cultural resonances with the high-tech geek culture of Silicon Valley.

Enquiring Minds draws from a slightly different repertoire of sources, based on a more sociological critique of the selection of curricular content according to social power, but it too endorses a collaborative inquiry pedagogy based on children's everyday cultures mediated through networked technologies. Its curriculum represents a crossover of children's cultures and school knowledge. These projects, along with others, reject the transmissive curriculum associated with academic content, the "visible" external products of learning, and graded student performance. Instead they advocate for a more acquisitive curriculum based on authentic experience and the "invisible" internal learning of the child.[25]

Enquiring Minds and Q2L are both grounded examples of curriculum futures being constructed through cross-sectoral and interorganizational networks. They embody soft networked governance in action. In turn, soft governance promotes different forms of knowledge. Putting it simply, the centralized curriculum as governed by hard government puts the stress on the centrality of the conservative canon as it passes this on from generation to generation. The decentralized curriculum governed through soft governance stresses diversity of learner experience, authentic contexts, and personal or collaborative inquiry— learner competence rather than cultural canon. Knowledge from different domains, everyday experience, and different cultural locations are all incorporated, criss-crossed, and interwoven into this vision of the curriculum of the future.

Governing the Curriculum

The switch from hard government to soft governance of the curriculum has begun to permit a greater diversity of players to participate in curriculum design. This changes the nature of the relationship or correspondence between schooling, economy and government. As shown in this chapter, the old model of schooling to work imagined curriculum as a direct mechanism for preparing students for work according to the needs of industry, and more recent work on school commercialism has shown how private-sector organizations have exerted influence on the curriculum. The turn to soft governance has now been shown to permit all sorts of agencies and relations across public- and private-sector interests as well as national borders and boundaries to participate in the actual construction and control of the curriculum.

These relationships and networks are lubricated by key intermediaries and crossover organizations and actors who crisscross traditional sectoral divides. The good ideas of these intermediaries are derived from an explosion of creativity discourse that is both linked to economic renewal—as embodied in cool, soulful capitalism and the affective playbor of the creative and digital industries—and to the everyday creative passions of young digital pioneers. The future of the economy is positioned as being dependent upon creativity and innovation that in turn are to be promoted and encouraged through new and innovative forms of schooling.

Soft governance, then, is not simply a new structurally flexible way of organizing education. It works into the very fabric of the politics and values of the curriculum, as the new soft style of "affect management" in the curriculum demonstrates. Governance is about a new way of organizing public education that involves the private sector and other intermediaries and crossover organizations and individuals doing parts of the work usually done by a central education system, and it does so through promoting partnerships with seemingly neutral intermediaries or through encouraging philanthropy. The curricula of the future described in this report, then, are interlocking parts of a complex and decentralized series of changes in public education that will see an erosion of boundaries between public and private sectors and the takeover of public functions by hybrid cross-sectoral and crossover actors and experts. The curriculum of the future embodies in microcosmic form how a more polycentric, multipolar education system, or centrifugal schooling, might work.

5 Psychotechnical Schools and the Future of Educational Expertise

Tracing cross-sectoral relationships in the fabrication of new curricular models is revealing, but it obscures a more subtle set of networks of relations that are facilitated by governance. These are the connections that are now brought about between concepts, ideas, visions, and the sources of expertise that promote them. As a result of soft governance, the curriculum of the future is subject to a proliferation of voices of apparent authority and the forms of specialist expertise, theories, ideas, philosophies, and other forms of knowledge they promote. Expertise creates the explanations and interventions that constitute a style of thought.

This chapter examines what sources of expertise and what professional knowledges are now being brought together as a style of thought for intervening in the makeup of the new curriculum. Two main expert groups now seem to be controlling the agenda for the curriculum of the future: psychologists and computer scientists. What style of thought do they deploy? What kind of politics and values are catalyzed by the deployment of their professional expertise in curriculum design?

The Expertise Explosion

The curriculum is the result of ongoing contests and negotiations over official knowledge. It relies on particular authorities to give it legitimacy. Yet the imaginary curriculum of the future articulated in many current curriculum projects depends on very diverse sorts of authority, much of it assembled as a messy juxtaposition of different voices. Authority over the concepts and principles of the curriculum has proliferated to include all sorts of dispersed sources and influences. This dispersal of authority is a micro-level refraction of social changes in the government and expertise of everyday life.

Sociologists have already begun to show that in many aspects of daily life we turn to highly diverse sources of independent authority and formally autonomous expert opinion. The way we think draws upon the expertise, vocabulary, theories, ideas, philosophies, and other forms of knowledge that are available and "speak to" us. According to many theorists, political and economic power elites no longer simply exert their governmental powers over the population by force or coercion. Instead, society is governed by a highly diverse network of institutions, programs, and techniques that have translated the needs of society into the personal concerns and mentalities of each individual of the entire population.[1]

In such a society we are all encouraged to take more responsibility for ourselves, and the human sciences provide much of the expertise we use to make sense of our everyday lives. Psychology, medicine, and economics are formal sources of such authority. More mundanely, self-help experts, diet experts, and money-saving experts help transport these authorities into daily life. Moreover, today it seems that there has been an explosion of expertise as the Internet has allowed formerly expert knowledges to escape formal professional control. The experiential expertise of "lay experts" generates and authorizes its own knowledge through Web communities that mediate professional expertise "at a distance." Schools, too, translate various voices of authority into programs and practices that work upon the minds and mentalities of the young.[2]

This means we need to be on the lookout for the "little experts" to whom authority is now increasingly accorded. These little experts, the experts of everyday experience, act as mediators who translate big ideas and styles of thought such as those of governments into the mundane and distant concerns, aims, anxieties, and aspirations of individuals. In economic life, for instance, the economic fates of people are understood as a function of their own particular levels of enterprise, skill, inventiveness, and flexibility. Consequently, each individual is solicited as a potential ally of economic success, and people are encouraged be "self-enterprising" and to invest in the management, presentation, promotion, and enhancement of their own economic capital as a "lifelong project." Through little experts, powerful capacities can work at a distance to align the objectives of authorities with the thoughts and aspirations of individuals.[3]

In professional activities like educational reform, the mediating role of the little expert is often played by a particular kind of policy specialist, or an intellectual worker. The intellectual worker is an enabler, fixer, catalyst, and broker of ideas, rather than a formal policymaker. Their ideas are "vehicular" or "propellant"—in other words, they move things on. Vehicular ideas are typically concerned with small-scale creative innovations, carried out in collaboration with a variety of constituencies and by means of a juxtaposition of people and ideas in order to bring about something new. Vehicular expertise, then, is not concerned with the grand schemes of big legislation but with practical, usable, marketable ideas capable of arousing attention and propelling the buzz of creativity and innovation. Such expertise contributes to a constantly mobile, creative culture of new ideas, new innovations, and intellectual creativity.[4]

In terms of curriculum reform, the mediator may be understood as taking big, abstract ideas such as the knowledge economy or globalization and turning them into practical programs. The mediator links the general to the particular and shifts a way of thinking from its original source of authority to a multitude of distant places. In so doing, the mediator juxtaposes certain kinds of expertise and seemingly authoritative ideas, bringing them together in order to get things moving. What kinds of professional expertise and intellectual creativity are associated with the design of the curriculum of the future?

Edu-Experts

These mediators are now importing new sources of authority and expert knowledges into public education. The new educational mediators are edu-experts who bring forth with them good ideas that envision and position schools and seek to "make up" the curriculum and construct learners and teachers as new kinds of individuals. The edu-experts behind new programs imagining the curriculum of the future speak a language of curriculum—worked through the rationale of innovation and creativity—that set parameters on and perimeters around the possibilities of classroom action and practice.

A surge of new vehicular ideas for innovative educational reforms have been propelled by the "policy intellectuals" of think tanks, NGOs, nonprofits, and other cross-sectoral agents and agencies. The "ivory tower" intellectual expertise of scholars and the formal authority of government-appointed experts have been increasingly marginalized as these new intellectuals cast about for ideas that seem as though they might work.[5]

The field of educational technology has been especially propelled by the ideas of a range of actors from across fields of education and learning, media, computer science, and from nonprofits, Web startups, and commercial R&D labs. Educational technology certainly has its gurus and talismanic leaders, but it is also a field constantly moving forward through the juxtaposition of new people and new ideas. Indeed, as already seen, the informal learning of young people in mediated environments is now regularly held up by these actors as itself a legitimate source of expertise for curriculum reform—a digital and youthful form of everyday experiential "lay expertise."

Curricula of the future are all the products of mediators, intellectual workers, and little edu-experts, who are propelling big abstract problems like globalization and technological change into the intellectual center of schooling. Formerly within the purview of the appointed experts of formal education systems, the curriculum of the future is now in the self-appointed expert hands of cross-sectoral intellectual workers who bring into the process of curriculum design a new set of techniques for getting things done and a new set of intellectual sources for thinking about the purposes and objectives of the curriculum.

Often these sources of expertise can be detected by tracing the intertextual links made by key curriculum texts with other external texts. A good example here is the Learning Futures report "Engaging Schools." In this publication, direct references are made to the US-based Partnership for 21st Skills (from which it cites the need for schools to promote new skills of collaboration, information literacy, and adaptability); a report on the "nature of learning" by the multilateral OECD; a think piece from a leading British conservative think tank and another from a think tank associated with the political opposition; a report from the British government schools inspection office; some key "meta-analyses" of research on cognition and technology from the fields of psychology, computer science, and the learning sciences; and a report on innovation in education commissioned by the multinational computing firm Cisco. In addition, the organizations behind Learning Futures have collaborated with High Tech High in San Diego to produce a guide to project-based learning.[6]

The actual references here are not as important as the expertise that they too rely on. Behind each of the references lies a repertoire of expert sources and selections from authoritative professional knowledge. This single example gives some indication of the ways in which the curriculum of the future is assembled from a messy and heterogeneous mix of references and authorities, each offering its unique expertise, enrolled from local, distant, and globally mobile sources. Fields of academic expertise, think tank opinion from different political perspectives, and corporate knowledge are all mixed together to constitute a depoliticized, cross-sectoral, transnational language of learning and innovation in the global dynamics of the twenty-first century.

The various originators of the new curriculum programs are mediators and little experts who catalyze and move their vehicular ideas through a variety of relationships. Behind many of these ideas lies the authority of a particular form of expertise. That is, the expertise of psychology. It is primarily by working through the ideas and expertise of psychology that the curriculum of the future can shape the minds, mentalities, and identities of students.

Making Up Minds

The new experts of curriculum reform are "little engineers of the human soul" rather than the "cold monster" of central government and departments of education. They are minor figures whose knowledge and practices seek to normalize particular ways of thinking, acting, and feeling in schools—to "make up" particular kinds of students. According to this theory, individuals are understood and "made up" as certain kinds of people through various kinds of knowledge and techniques. Schools and curricula act as apparatuses that accord to students all kinds of new possibilities of perception, motivation, emotion, self-reflection, and so on. Little experts in the field of education, then, have the power to shape how we understand students as people with particular competencies and capacities, a task they have done through the knowledge, authority, and techniques of psychological expertise.[7]

These little engineers of the human soul bring with them into the imagining of the curriculum a particular conception of the schoolchild (or an idea about who the schooled child should be), along with interventions to act upon them. These curricula construct the child not as a passive recipient but as an active producer of knowledge. Q2L, Enquiring Minds, New Basics, and the rest all talk about active learners and knowledge producers in dynamic systems, relations, and communities of practice. Recall that High Tech High refers to its organizational "soul." What is at stake here is a reengineering of both organizational souls and human souls. Through expertise tne values and goals of the educational organization, and the authorities on which they rely, are brought into contact with the dreams and actions of children. Political, social, and institutional goals are aligned with individual pleasures, desire, and happiness. Contemporary curriculum design works through the deep inner soul, interior life, and habits of mind—the emotional and affective state of personal development—of the whole child understood as an "active learner," a "constructivist learner," and an "autonomous learner," and it mobilizes appropriate "interactive pedagogies":

In interactive pedagogy . . . the teacher teaches by adapting the material to the child's momentary interests and imparts information that is set by the children's questions. This pedagogy requires the teacher to respond flexibly to the child's feelings, words, and actions. . . . Interactionism constructs both a response-able/-ready child and a response-able/-ready teacher. . . . Interactionism . . . can be characterized as fluid, dynamic, situation responsive, pragmatic and virtual.[8]

Flexible interactive pedagogies that respond to the dreams and actions of the child are now the preferred pedagogies of the curriculum of the future, as defined by the expertise of a host of little experts and engineers of the human soul.

The emphasis on "inquiry" in many prototypical curricula of the future is a good example of the new pedagogical expertise of the soul. Inquiry is a very particular form of knowing. It has its roots as an educational concept at least as far back as Dewey but has attained particular significance as a way of knowing in a dynamic networked era. Futurelab's Enquiring Minds emphasizes inquiry as a way of knowing that is necessary in a complex informational environment where it is more important to know how to seek and how to analyze information than to acquire and retain basic knowledge. The task for teachers in an inquiry classroom is to listen and respond to students, adapting flexibly and fluidly to their interests and questions accordingly. Inquiry learning and interactionist pedagogies are mutually interdependent.

Quest to Learn and Learning Futures have generated the same kind of pedagogies. Learning Futures views inquiry as research, experimentation, problem solving, and evaluating information, while Q2L's "evidence-based inquiry curriculum" is modeled to drop learners into "inquiry-based, complex problem spaces that are scaffolded to deliver just-in-time learning." The capacity for inquiry is not, though, a natural and latent part of the character of students, just waiting to be set free once the conventional curriculum has been cast off. Inquiry, like creativity, needs to be promoted, encouraged, managed, and finessed, and the ideal pedagogy for accomplishing this is a responsive form of interactionism. Students need to be made to be inquiring. These projects are all, it seems, involved in making up inquiring minds.

Inquiry is part of a more wide-reaching discourse of "competence" based on the invisible, internal learning of the child. As noted earlier, competence is constituted through the discourse of active learning and creativity; self-regulating learners; a pedagogic discourse of interactivity, projects, themes, and experience; learner autonomy over the selection, sequencing, and pacing of learning; and the intentions, dispositions, relations, and reflexivity of learners. Competence refers to the open narratives and personal projects of the individual, their cognitive, affective, and motivational dimensions, rather than to the grand collective narratives of the disciplines that make up the subject-based curriculum.[9] Competence, in other words, is the technical descriptor for the child's soul.

Creativity with Attitude!

The highest possible form of competence seems to be creativity. Creativity is important because it is both a human capacity—we are all more or less creative now—and an economic imperative. Psychologists from various subdisciplines have been highly active at promoting creativity both as part of everyday psychological life and part of an "entrepreneurialization of business and economic life."[10]

To take one example, texts like those produced by the self-appointed experts of the British think tank Demos have repeatedly sought to "realize the creative potential of all citizens and to boost competitiveness in the knowledge economy" by making "radical changes to the education system."[11] The "creative age" imagined by Demos is a "radically high-tech, corporate democracy" in which "creativity with attitude" is described in the same terms found in "creative management" and "self-help manuals on 'creative thinking' and 'creative living.'"[12]

Creativity now spans academic, popular, personal, political, educational, and business spheres, and schoolchildren are positioned as inquiring, competent, "creative souls" whose inner lives and habits of mind—defined psychologically—are to be the subject of interactionist pedagogic intervention.

The Whole Education network is part of the shockwave of the creativity explosion as it has been felt beyond the enclosures of the psychological disciplines. Based on a wide array of sources of authority and expertise, Whole Education constitutes a network that is bound together loosely by a series of "common beliefs." These common beliefs intertwine creativity, employability, and personal competence. Whole Education promotes "adaptable and creative" learning "throughout life"; "independence" and the development of "every individual" through a "diversity and choice of education pathways"; "building resilience" and "teaching social and emotional competencies including self-awareness, empathy, self-respect, persistence, and self-discipline"; forging "strong relationships" and "collaborations"; taking "joint responsibility" and practicing "active citizenship," and supporting learning "outside school, in the community and online."[13]

Embedded in the ideas of inquiry, competence, and creativity then is the extension of a largely psychological way of understanding and working with students. Competent and creative inquiry looks introspectively; it is concerned with students understood psychologically, cognitively, and affectively rather than those understood sociologically in terms of social structures, knowledge, and collective narratives. The role of teachers is to interact with students in order to facilitate their competence. Competence puts the onus on self-understanding and self-fulfillment, as shown in the stress put on creativity and its correlates of learning to learn, constructivism, metacognition, effective lifelong learning skills, multiple intelligences, and so on, which position learners as inwardly focused private souls.

Psychotechnical Schools

The construction of greater synergy between technology and the curriculum—symbolized by the emphasis given to inquiry, competence, and creativity rather than knowledge—means that all of these elements are now becoming part of a new psychological way of managing the curriculum. The future of the curriculum is subject to a new form of professional psychological expertise that acts to shape students as creative souls through reshaping curriculum. The curriculum embodies learning how to see, think, feel and act; it shapes identities and minds. In the psychological management of the curriculum, the perspectives of psychology ("psychological eyes") generate the standards and rules by which students are to view themselves and participate in school while psychological concepts accordingly generate the principles and classifications by which the curriculum is to be reimagined and redesigned.[14]

The strength of psychological discourse—or "psy" for short-hand—in contemporary education is part of a long history of a whole complex of "psy disciplines" and their role in "making up people" as "inner-focused persons" through school. The "psy complex" consists of heterogeneous knowledge, forms of authority, and practical techniques that make up psychological expertise and the eyes or "gaze of the psychologist." Today, various forms and subdisciplines of psychology see the individual as an autonomous individual enmeshed in a network of dynamic relations with others. It is through the gaze of such dynamic and social psychologies that the contemporary psy complex operates. Through dynamic psychological expertise, psy promotes new styles of thinking about ourselves and others, our feelings, our hopes, our ambitions and anxieties, and new ways of planning life and approaching life's predicaments, realizing one's potential, gaining happiness, and achieving autonomy. We're "made up" as ideally and potentially a certain sort of psychologically autonomous person. The individual is viewed by these psychological eyes as an "actively responsible self" whose own personal psychological fulfillment and quality of life is allied to the achievement of wider political and economic purposes and objectives.[15]

Through its implantation in schools, the psy complex has made the learner the object of scientific know-how and therefore knowable as a subject of intervention in order to bring about a change in the future. Psy expertise has provided particular ways of thinking about childhood and new ways of seeing children that have spread to schools through a huge variety of texts, techniques, and practices that now make it possible to act upon their competencies and capacities in classrooms. The curriculum of the future applies dynamic psychological expertise, which sees young people enmeshed in networks of relations, to the problems of education in the digital age.

The psychological emphasis in education is nothing new of course; only now, however, it has been rearticulated in terms of its measurable economic contribution. An interesting example of this new alliance of inner focus and economic purpose is the Apps for Good program. As the project Web site describes it: "Apps for Good is an award-winning course where young people learn to create imaginative mobile apps that change their world. Our students create apps that make a difference and solve real life issues that matter to them and their community, giving them a launchpad in social enterprise and the exciting world of technology, design and innovation."[16]

The Apps for Good course links the creation of mobile apps to a philanthropic sense of purpose while also seeking to build students' "self-confidence" and readying them for "employment, self-employment and entrepreneurship in the real world." The psychological management of the curriculum constitutes a hybrid discourse that is simultaneously technological, philanthropic, psychological, and entrepreneurial. It is affective and focuses on feelings and passions, part of the participative culture of playbor noted earlier.

The hybridization of inner-directed psy discourse with economic entrepreneurialism is proliferating. For example, a British

research project synthesized a very large number of different "skills frameworks" emerging from government departments, research institutes, private companies, and crossover
or "third sector" organizations. It compiled a report on the
"wider skills" required for twenty-first-century economies. Its
findings emphasize the importance of "new smarts," "orienta-
tions," "capabilities" and "capacities," "dispositions" to learning, and the "mental and emotional habits of mind" that are
required "if innovation is to be effectively developed in young
people."[17] Another British research project identified very similar trends in an analysis of "personal skills and competences"
frameworks, while a third report stressed the strong connection
between improving personal "well-being" and "happiness"
through education and the enhancement of economic well-
being—a combination the report describes as a perfect "state
of happiness."[18]

The "wider skills" report proposes the application of psychological expertise to the economic challenges of the twenty-first
century and identifies methods for cultivating, tracking, and
measuring the new desirable qualities of "innovation." Young
people are positioned by the report as the subjects of psychological discourses of cognitive competence, emotional resilience, and therapeutic self-reflection. The stress on competence
is then couched in terms of how schools can cultivate the habits of mind that underpin innovation. Schools are encouraged
to promote a more active, creative, and innovative learner, in
order to ensure a more active, creative, and innovative future for
the economy. The well-being report echoes these conclusions.
The curriculum of the future, from this perspective, is concerned
with techniques to intervene in the psychology of students in
order to maximize innovation. These three reports constitute

and contribute to a discourse for the psychological management of the curriculum that is at the same time human-focused,
economically innovative, and seemingly politically progressive.
They amalgamate theories of competence originating in the lib-
eration of individuals' active creativity in the 1960s and 1970s
with emerging twenty-first-century psy theories of creative intelligence and the "new smarts" associated with innovation in a
knowledge society.

"Psychotechnics" was the name given to projects that sought
to intervene in the psychology of factory workers in the early
twentieth century. Psychotechnical projects were the psychological sibling to the hardline factory management techniques
of scientific Taylorism. Like the Taylorist techniques that sought
to ensure maximum efficiency on the factory production line,
psychotechnics sought to maximize the utility of the factory
worker by redesigning the work process and by sorting, select-
ing, and allocating workers to tasks on the basis of matching
their competence to the demands of the activity. Psychotechnics
sought to improve the "productive machine" by investing in the
"human machine" as an active, autonomous, and motivated
individual carrying out meaningful tasks. In today's culture of
playbor, work has become as much psychological as economic,
perhaps more to do with the identity of the employee than labor
and cash.[19]

Schools, likewise, are sites where psychotechnical ideas, procedures, and techniques have been employed in order to assess
and classify and act upon the capacities and competences of individuals in relation to political ideals and economic problems.
The world today certainly has its economic problems, and much
discourse on the curriculum of the future appears to advocate for
increased intervention in the competence and well-being of the

student in order to improve a nation's capacity for innovation, as firmly demonstrated by the "new smarts" report. This is close to the reality of how business processes and job descriptions are linked to electronic databases of individual competence profiles, based on human capital metrics, to align people with corporate objectives.[20]

The various competencies frameworks analyzed in the "new smarts" and "happiness" reports—and those embodied in projects like Opening Minds, Enquiring Minds, Q2L,and so on—may therefore be understood as psychotechnologies that act to make up learners in terms of competencies and capacities of flexibility, adaptability, initiative, ad hoc groupings, informality, innovation, and creativity. They are linked to a "new image of work" and a "new image of the worker," generated by psychological expertise, in which action, innovation, entrepreneurship, excellence, initiative, and so on can be released through the promotion of human autonomy, values, experimentation, creativity, risk, and innovation.[21]

Understood in this way, programs like Learning Futures and Quest to Learn seem to be advocating for a new kind of psychotechnical curriculum of the future. Instead of "human machine" metaphors, references to "well-being," with human well-being now fused to productive well-being, proliferate in these projects. In the psychological management of the curriculum, the machine itself has been humanized!

The key issue is that the psychotechnic projects of the curriculum of the future are not merely psychological. The psy discourse has now been blended with computer science in the transdisciplinary field of learning sciences to produce a hybrid of psy and CompSci styles of thought—a new "CompPsy complex" that merges psychological and computational thinking.

Through its amalgamation of psychological learning science and computer science theories and vocabularies, the CompPsy complex mobilizes a distinctive style of thought through a raft of new terms, concepts, references, arguments, explanations, and practical techniques of intervention. The learning sciences mobilize theoretical and philosophical descriptions from cognitivist, constructivist, constructionist, and sociocultural perspectives, augmented by computer, systems, and design sciences (and increasingly neuroscience too). The learning sciences provide detailed accounts of the technical and social processes of learning with digital technology, including its socially collaborative nature, but tend not to examine the social, political, economic, cultural, and historical contexts within which educational technology use takes place.[22]

The CompPsy complex is an emerging scientific field and style of thought, then, which melds understandings of the technical and immediate social contexts of learning with the design of effective interactive technologies, informed by computational thinking, and the psychological management of student emotions. It embodies certain values, concerns, and politics, and through the design of specific curricular programs and technical systems it catalyzes certain actions and experiences. Captured in the term "socio-technical change" used by sociologist of science and technology, technologies are outgrowths of social actions that carry with them a host of political associations and historical connections that they implant in human behavior, thought, and action through privileging certain activities, states of being, and positions over others.[23] The design of educational technologies by learning scientists has been described as a method for "designing people" through "engineering" particular forms of learning, actions, and dispositions.[24]

The style of thought of the CompPsy complex, then, generates certain sorts of experiences in the curriculum of the future, and it catalyzes certain sorts of pedagogies and interactions among educators and learners. In the discourse of CompPsy, authority is given to transdisciplinary knowledge, to innovation, and to creativity in addition to self-improvement, well-being, and personal competence. The objective of the CompPsy complex is to maximize human well-being, happiness, and self-competence while also seeking to maximize productive creativity and innovation for a high-tech global competition. Mental and economic well-being are mutually constitutive. It produces an ideal-type learner identity of the "individual entrepreneur" with "ethical-economic and psychological quality."[25]

The emergent CompPsy complex has now begun to exert its transdisciplinary scientific expertise on the shaping of the curriculum of the future, thanks to the network mode of governance and the diverse authorities it has permitted into curriculum design. The CompPsy complex seeks to act upon and make up persons to be self-managing in order to benefit an economy that requires expertise across informational and technical disciplines. In short, by ushering into the educational field a host of new crossover players, actors, and voices of CompPsy expertise, governance generates a particular kind of self-competent, inner-focused individual, an individual whose emotional well-being is important for innovation and the future well-being of the economy.

In summary, the replacement of an official canon of curricular knowledge with a new expertise of creativity and competence, and their classroom correlates of inquiry learning and interactive pedagogy, has been largely led by expert individuals and organizations whose links with national and state education systems are informal, loose, and shifting. Think tanks, NGOs, nonprofits, foundations, professional societies, and commercial networks, including those concerned with industrial modernization, enterprise, and the future of work in the digital age, have become the self-appointed little experts of the curriculum for the future.

The transdisciplinary blend of psychological and learning sciences approaches to education advocated by these intellectual experts has sought to position students as inner-focused individuals whose own self-responsibility, competence, and well-being—their deep inner soul, interior life, and habits of mind—have been fused to the political objective of economic innovation. Their own self-fulfillment, mental and emotional well-being, and happiness are important for global economic well-being. Education is important in this respect. Rather than being "schooled to work" as "human machines" assigned to be components of the "productive machine," an emerging CompPsy complex assigns human well-being to productive well-being through psychotechnical visions of the future of schooling. In this sense, the new psychotechnical edu-expertise has fused educational effectiveness to the more affective realm of culture.

6 Globalizing Cultures of Lifelong Learning

Although a curriculum is often allied to political and economic objectives, it is also linked to culture. In the recent history of the curriculum, a conservative version of culture has predominated. Schools have been charged with communicating great cultural works, a largely Western-centric version of history and geography, and a canon of scientific knowledge. Alongside the official curriculum lies a "hidden curriculum" that stresses, among other things, the traditional values of family, elite culture, patriotism, and capitalist economics. All of this contributes to what schoolchildren see as "real" and important. As a series of selections from culture, a curriculum is a message about the future embedded in a particular vision of what real culture ought to be.

Any curriculum of the future is therefore involved in establishing what may be seen as real culture in the future. At the present time, many prototypical curricula are seeking to establish the culture of the Internet as part of the legitimate culture articulated via school. What kind of selection from Internet culture, therefore, is being worked into the curriculum, and what cultural visions and values for the future are being established as a "reality" for schoolchildren?

Global Cultural Patterns

According to studies of culture and communication in the age of the Internet, we now inhabit a global communicative universe that is multimodal, multichannel, and multiplatform. Mass media such as TV and newspapers have converged with personal communication in the new cultural landscape of social media, bringing about a more participatory form of culture (rather than passive spectatorship) where consumers are encouraged to seek out information and make connections among dispersed media content.[1]

The convergence of old and new media has given rise to a new form of mass communication, or "mass self-communication" that prioritizes "my time" over "prime time." In the universe of Facebook, YouTube, and so forth, people are now enabled to communicate and interact on a previously unimaginable scale as "creative audiences." However, the massive potential of creative audiences to reshape, reproduce, and recirculate media—or to produce original content—is shaped and controlled by a concentration of interlocking corporate multimedia, financial trade, and government strategies that have permitted the expansion of for-profit entertainment and the commodification of personal freedom.[2]

In a convergent media culture, then, we see both a greater degree of control and creativity among audiences and consumers, and a greater concentration of ownership and commodification among commercial media producers. It's not simply a case of grassroots bottom-up media and the free culture of hackers winning over the top-down mass cultural model of the corporate high-rise, but of how they engage in complex conflicts and struggles, or conversely how they reinforce and reward one another. The interactions of creative audiences and commercial producers today are shaping the future of Internet culture specifically and popular culture more generally. The result of convergence has included the emergence of four interacting cultural patterns. The first two are communal and the latter two are individualist: (1) cosmopolitanism: greater opportunities for engagement with global causes; (2) multicultural hybridization: the global remix and circulation of diverse (multi-) cultural products from around the world; (3) consumerism: the formation of a global capitalist market based primarily on branding in a commodified culture; and finally, (4) networked individualism: the construction of individual cultural worlds in terms of personal preferences and projects. Networked individualism is a culture that starts with the values and projects of the individual who interacts with others following their own choices, values, and interests, rather than by tradition and hierarchy. Networked individualism is the most prominent cultural pattern of the Internet:

The culture of networked individualism finds its platform of choice in the diverse universe of mass self-communication: the internet, wireless communication, online games, and digital networks of cultural production, remixing and distribution. . . . The culture of networked individualism can find its best form of expression in a communication system characterized by autonomy, horizontal networking, interactivity, and the recombination of content under the initiative of the individual and his/her networks.[3]

The culture of networked individualism is not just selfish individualism. It can inspire social movements, based on the sharing of new cosmopolitan and multicultural values, that may become insurgent communities of practice. Networked individualism can also lead individuals to entrench themselves in the already-constructed values and branded identities of consumer-media culture.

Although the Internet as a medium itself can also diffuse cosmopolitan, multicultural, and consumerist values, it is important to reiterate that the "cultural roots of the Internet" have been traced in "the culture of freedom and in the specific culture of hackers." A "cultural resonance" has therefore been established between the culture of the designers of the Internet and the rise of a culture of networked individualism and creative audiences that finds its way into the minds of millions of Internet users. Networked individualism, with its focus on personal choice, projects, and self-entrepreneurial behavior, is the globalized cultural expression of a set of Silicon Valley cyberlibertarian values.[4] Geek politics have gone global!

In other words, the cultural roots of the Internet now resonate through the popular culture of the Web. As Internet culture is increasingly directed into the curriculum of the future, a cultural resonance may be established between the Internet and education too. The consequence, it seems, is that the curriculum of the future is to be programmed according to the cultural aspirations of networked individualism and an emphasis on personal choice, personal projects, and self-enterprise implanted in Internet culture by the computer engineers and "geeks" of Silicon Valley. Does this mean that the geek politics of Silicon Valley has been embedded in the curriculum? How to design a curriculum to respond to the globalized cultural patterns of the Internet is now a key issue.

New! New! New!

The New Basics project in Queensland, Australia, emphasized cultural globalization as a context and a rationale for curriculum reform. The main text for teachers generated by the project team stated: "The New Basics are futures-oriented categories for organizing curriculum. Essentially they are a way of managing the enormous increase in information that is now available as a result of globalization and the rapid change in the economic, social and cultural dimensions of our existence."[5] The New Basics stressed a series of transdisciplinary curriculum categories, each framed by a question. These categories and their questions were: life pathways and social futures (who am I and where am I going?); multiliteracies and communications media (how do I make sense of and communicate with the world?); active citizenship (what are my rights and responsibilities in communities, cultures, and economies?); and environment and technologies (how do I describe, analyze, and shape the world around me?).

The New Basics is a clear example of a curricular response to the perceived changes of cultural globalization. It considers the curriculum as a selection or allocation of values, and recognizes that globalization has challenged the sorts of values that are to be imparted and reproduced by any curriculum. At the same time, however, the rather progressive focus on life pathways and active citizenship subtly reframes the more instrumentalist concern of how to shape workers for the competitive pressures of economic globalization.

As one study of the New Basics phrased it, the title "New Basics" appealed to a cross-section of the educational community, from progressives who liked the notion of the "new," to conservatives who liked its "basics." The project documentation is full of references to the "new." It mentions "new student identities," "new workplaces," "new technologies," "new times," "new citizenship," "new knowledges," and "new epistemologies" in order to construct its futures-oriented curriculum.[6]

The discursive hybridity of conservative and progressive ways of thinking about curriculum captured by the title "New Basics" is continued in the thematic curriculum organizers. Weight is given to the importance of diverse family relationships, interaction with local and global communities, local and global economic forces, the historical foundations of social movements and civic institutions, developing a scientific understanding of the world, and working with design and engineering technologies. In these categories, family, locality, history, civic institutions, and scientific understanding are established as the basics or the foundations to which the new demands of diversity, global communities, global forces, and new technologies must now be added.

In the version of globalization constructed by the New Basics, a very cosmopolitan vision of curriculum is required. Cosmopolitanism represents the sharing of values on a global scale that transcend local and parochial interests. Such concerns are linked to the diversity of multiculturalism, changes in traditional family structure and everyday family life; to the expansion of notions of community and civic participation, powered by digital media, and its effect on the individual's capacity for belonging; as well as to global economic and political forces.[7]

Besides attempting to reform the curriculum in order to develop the skills and dispositions perceived to be required by the knowledge economy and globalization, the New Basics curriculum is part of an attempt to reimagine community in the context of multiculturalism, global cultural cosmopolitanism, and the pressures these shifts have exerted on the national community. To an extent, then, the New Basics may be seen as a curricular extension of the major cultural patterns of cosmopolitanism and multicultural hybridization in a global network society.

Everyday (Media) Cultures

The Enquiring Minds curriculum R&D project run by Futurelab in the United Kingdom also sought to address a changing perception of community in the context of cultural globalization. At the root of the project was an interest in the various communities now understood to constitute children's everyday cultural experiences. As the main curriculum guide documentation states, Enquiring Minds was not so much concerned with the improvement of pedagogy or with students' learning processes but with "the relationship between this and what they are learning," and it was intended to "explore the potential for students' own experiences, interests, concerns and lives to act as the starting point for creating a meaningful, relevant and engaging curriculum for young people. What has been ignored in debates on the development of effective pedagogy has been the question of how learning is intimately tied up with the question of knowledge, or of how we address the questions: learning what? for whom? and why?" The EM guide states that "the relationship between pedagogy and curriculum and between 'school' knowledge and students' 'informal' knowledge is central to the search for more effective and powerful educational strategies for the 21st century."[8]

In response to this challenge, EM offers a view of a possible future curriculum that puts everyday culture at the heart of the curriculum enterprise. It draws, again, on the radical pedagogy of Paolo Freire and a sociological explanation for the curriculum. The project recognizes that different curricular formats are produced by different configurations of social power that seek to produce different student mentalities, with academic bodies of knowledge embodying mentalities that are intellectual, abstract, and active while practical and vocational pedagogies may be associated with more concrete and passive mentalities. That is to say, different students' mentalities are built into the deep structure of the curriculum form.[9]

Enquiring Minds offered a curriculum format that "de-differentiated" students' school knowledge from their everyday or informal knowledge. It stressed students working with cultural knowledge—understandings and meanings related to specific events and objects—and with critical knowledge that would allow them to understand and critique the forces that shaped the world. Instead of fixed school knowledge, it advocated for "dynamic knowledge" to be the subject of a reinvigorated future curriculum. Dynamic knowledge is open to change; it is recognized as constantly in production, often contested, socially contextual, and transformed in reality. The EM guide stresses that "the development of the curriculum starts with students' interests, ideas and experiences," and that the task for teachers is to help them "explain, expand and explore further from that starting point . . . to illuminate or decode aspects of their experience."[10]

EM sought to promote a curriculum form that saw students' everyday knowledge and cultures as worthy of attention in the curriculum. Rather than setting up students' concrete cultural experiences as inferior to the reified knowledge of the formal curriculum, it understood culture itself to be a complex site of human activity in which knowledge is shaped, produced, and revised over time. It additionally saw students themselves as actors who, through a range of critical pedagogies and inquiry-based techniques and practices, might themselves shape, produce, and revise cultural knowledge by utilizing the "building blocks" of ideas and concepts from a range and blend of subjects.

Moreover, EM acknowledged that young people's uses of digital media and technology offered a challenge to the curriculum. The approach of EM, however, was not to advocate for the kind of skills and competences that were earlier associated with a series of "soft openings" in curriculum reform. Instead, the rhetoric of EM constructed "the informal curriculum taught through media and leisure" as itself problematic, as the EM curriculum guide detailed:

Media corporations have figured out their own 'pedagogies' and become modern society's best teachers. The corporate curriculum of consumer culture has, in turn, become a yardstick against which the school curriculum and its associated pedagogies are assessed. . . . However, consumer-media culture teaches particular sorts of knowledge, and these are based on affective pleasures rather than the more reflexive pleasures of knowing about and being able to interpret the world. Being a media consumer is one thing; being an informed and critical consumer is another.[11]

Pretty explicitly, EM offers a construction of a curriculum as a critical pedagogy of consumer culture intended to promote student mentalities of critique. It provided a response to the cultural pattern of branded consumerism.

However, in its cultural emphasis, EM also implicitly advocates for the curriculum as something that is both learned in school and out of school, lifelong and lifewide. Here the complexities of linking curriculum and culture are most clearly seen, because lifelong learning may be largely understood as itself the dominant informal curriculum form of consumer-media culture.

Lifelong Learning

A review of alternative curricula carried out by Futurelab at the same time as the Enquiring Minds project showed how projects and portfolios have become an essential pedagogical component in curriculum reform. Many new curricula include an "extended project" or "personal challenge" component that is seen as a means of ensuring that learning is meaningful and coherent, enabling development of learner responsibility and allowing learners to develop skills and competencies that could not be developed through other pedagogic approaches. Such personal challenges are characterized in the review documentation as "content-neutral," as taking place in "authentic contexts," as making a "contribution," and as enabling learners to "make connections across different subject areas and across in-school and out-of-school learning" supported by "specialists across and outside the school community": the boundaries between "specialist subjects" and "specialised areas of personal interest" are routinely punctured.[12]

Almost all of the prototype curricula gathered under the loose umbrella term "centrifugal schooling" feature a project-based element. Learning Futures, High Tech High, Enquiring Minds, and Quest to Learn all emphasize student inquiry through focused project-based learning. A similar model is that of "rich tasks" derived by the New Basics. Rich tasks are not short-term projects but problems that require "identification, analysis, and resolution, and require students to analyze, theorize and engage intellectually with the world" outside the classroom through transdisciplinary practice.

A document produced in a collaboration between the organizers of the Learning Futures and High Tech High programs outlines guidance for teachers in promoting extended, interdisciplinary project-based learning, which it describes as "designing, planning and carrying out an extended project that produces a publicly-exhibited output such as a product, publication or presentation." Moreover, it claims, "digital technology makes it easier than ever before to conduct serious research, produce high-quality work" and to "foster a wide range of skills (such as time management, collaboration and problem-solving) that students will need at college, university, and in the workplace." The text constructs project-based learning as a pedagogy that transcends classrooms and prepares students for all walks of life.[13]

The project pedagogies put forward in these programs can all be viewed as part of the same broad societal emphasis on preparing students for lifelong learning. Lifelong learning is here understood as the dominant pedagogy of a futuristic "learning society" in which learning is not narrowly canalized by a few educational institutions but dispersed diffusely into the very atmosphere of society. A learning society is both a planned society, driven by the need for governments to ensure their people are constantly equipped with the occupational competencies required to remain competitive, and a reflexive society. A reflexive society implies the capacity for everybody to learn new things in order to keep abreast of very rapid societal change in which the knowledge they acquire is no longer certain and established forever. Being reflexive means being constantly self-examining and having the ability to adapt one's own behavior to changed conditions and innovations. Learning in such a society is therefore a whole way of life that is continuous and nonstop.[14]

For example, High Tech High and Learning Futures both put the emphasis on learners producing an ongoing digital project portfolio, making links between their own out-of-school interests and the needs of their communities with the curriculum, and on preparation for the adult world. Rather than putting the stress on acquiring knowledge, the HTH curriculum stresses the development of a preferred model of adulthood as its outcome.

The active, self-directed pedagogy of the lifelong project has also been idealized by research on online learning and the participatory cultures of the Web. "Shape-shifting portfolio people" who think and act in terms of their résumé, and who define their own personal projects in entrepreneurial terms as businesses or enterprises, have been imagined as ideal-type flexible, interactive, and constructivist learners able to continue learning and adapting, based on constant reflexive self-analysis, right through the life cycle.[15]

The personal challenge or project is the ideal pedagogic mode to promote the ability to be taught, continuously and lifelong, across school and out-of-school communities, throughout a "pedagogized future." The emphasis on continuous learning is captured in the idea of a "total pedagogy," which means a continuous disposition to be trained for the requirements of an entire life in a process that is permanently open.[16] The shape-shifting portfolio person is the perfect figure for a permanently open, totally pedagogized future. For many critics, though, the kind of pedagogized futures most young people can expect are also highly consumerized futures.

Consumer-Media Curriculum

Consumerization refers to the process of becoming increasingly consumerist, the growth of consumerism, and the action of making something more appealing to consumers. To speak of the consumerization of learning therefore recognizes that learning itself has become both increasingly consumerist and more appealing to consumers. The market is taken to be an educator in itself.

The "market-as-educator" approach argues that the commercial market of computers, TV, toys, and popular culture teaches children in informal ways that appear to "clash" with what they can expect from teachers and formal education. Children's existing consumer-media cultures have been identified as rich and seductive learning environments in their own right; thus a competition has been established between the competing resources of the global corporate curriculum of consumer-media culture and that of schooling. Commercial organizations, it is said, have been better than education systems at aligning themselves with the lifestyles, identities, and ego-projects of young people who seek to identify themselves as autonomous, pleasure-seeking consumers.[17]

Put even more critically, it has been claimed that today "the curriculum of our culture, 24 hours a day, 7 days a week, 365 days a year, is advertising." This cultural curriculum of advertising seemingly allows "corporations [to] deliver a broader ideological message promoting consumption as the primary source of well-being and happiness," and it positions young people less as "active citizens-to-be" and more as "passive consumers-to-be."[18]

Consequently, consumerism, commercial life, and the world of goods have been "naturalized" as a seemingly benign aspect of children's lives. Children are not as much brought into consumerism by adults, whether by caring parents or teachers, or seduced into it by media and marketers, as born into it through commercialized parenting pedagogies. Consumption is a lifelong activity with the life course itself commodified in relation to commercial interests, practices, and processes.[19]

Digital media are a significant source of the consumerization of learning. As digital media have become more sophisticated and increasingly accessible, the range of learning options catering to all tastes and interests, now waiting to be consumed, has proliferated. Learning activities have become consumer goods in themselves, purchased within a marketplace where learning products compete with those of leisure and entertainment. In the culture of consumerized learning, learning is central to lifestyle practices. The consumer needs to be always learning about new lifestyles. Consuming is learning and learning is consuming. Lifelong learning is now to become lifelong lifestyle learning.[20]

The point to make here is not that lifelong learning and project-based learning are somehow linked to consumerism or to the consumerization of the curriculum. It is to stress the importance given to lifelong "projects" as a cultural pattern of networked individualism. The personal project has become a new and continually ongoing state of mind in a "cut-and-paste curriculum" orientated by individual self-responsibility, personalization, and technology-based child-centeredness, with students encouraged to make "a planning office for themselves." Likewise, in the culture of lifelong learning, learners are to make projects for themselves in order to express their "educated" anxieties and aspirations. Through the language and practice of projects, young people are being sculpted and molded as malleable, shape-shifting, lifelong learners with the competence and capacity to be autonomous, self-responsible, and self-enterprising in both their choices about lifestyle and learning.[21]

Nowhere is the shape-shifting potential and networked individualism of learning more forcefully advocated than in the resources of the Web. By shifting learning outside of the school gates, and setting it free in a cultural landscape rich with multimedia, the practices of learning are hyperlinked to the curriculum of commercial culture, a culture that for some educational commentators is participatory and sophisticated yet for others ideologically regressive and aggressively commercialized, connected to highly ideological ideals of free market education without any intervention from the state.[22]

Although Enquiring Minds assembles something of a critique of consumer culture into its curriculum framework, the researchers were left at the end of the project wondering if it had achieved anything emancipatory or simply enmeshed students more firmly into the consumerized contexts of their everyday cultures.[23] The difficulty encountered by the project has been in differentiating its discourse of child-centered inquiry and personal projects from the individualism associated with both the political right and with the networked individualism of personal autonomy most clearly found in the culture of consumerism. The discourse of Enquiring Minds is one of freedom and choice, terms that resonate with the cyberlibertarian, entrepreneurial culture of networked individualism and the market-as-educator culture of active consumption. In all, the individual is expected to pursue their own separate and autonomous development, to manage their projects and their portfolios. Their identities are being sculpted by a particular style of cultural thought that emphasizes concepts of do-it-yourself (DIY) self-shaping.

7 Making Up DIY Learner Identities

This chapter centers on the issue of how a curriculum translates ideas about who students are and who they should be. The curriculum promotes and sculpts learners' identities, their minds and mentalities. What you know makes you who you are. Learning the content of the curriculum is not simply about acquiring and understanding school knowledge. It embodies learning how to think, feel, and act as certain kinds students and as certain kinds of people. The curriculum of the future, as we have seen, is the product of a style of thought that draws on concepts and references regarding knowledge, networks, the economy, psychotechnical expertise, and the cultural patterns of globalization. What identities are promoted and molded by the style of thought underpinning the design of the curriculum of the future? What will the students of the future learn as appropriate ways of thinking, feeling, and acting? According to what future aspirations and objectives as described by what authorities?

As previous chapters have shown, the curriculum of the future is a hybrid of new learning languages, technological systems and network-based discourses, new links with the economy and discourses generated by governance, and cultural discourses of globalization. Key elements of the discourse of centrifugal schooling and the curriculum of the future include networked and connected learning, psychological competence in inquiry and creativity, and the ability to make one's own projects as a lifelong endeavor. The identity promoted by this amalgamation of elements is that of a "DIY networked individual."

Prospective Identities

The curriculum is never simply a matter of passing on information from one generation to the next. It embodies learning how to see, think, feel and act. It shapes identities and mentalities. The construction of any curriculum therefore implies the making of kinds of people. It invents and promotes preferred kinds of identities and mentalities that, through ongoing study, students are encouraged to adopt as their own schooled identities. The emphasis on "human capital" for the economy, for example, is a clear case of purposeful identity formation.[1]

In the case of the traditional conservative-restorationist curriculum, the kind of content that is taught stresses the importance of the past, as embodied in cultural canons, the ideal of universal knowledge, and so forth. The curriculum of the past promotes a "retrospective identity" through narratives of the past; through such identities it is hoped that the narratives of the past will be conserved and projected into the future. That is, students will carry on these narratives of the past into the future in their own mentalities and identities—the ways they see, think, feel, and act—and project them into their own aspirations. The curriculum of the future, however, promotes "prospective identities" that are "constructed to deal with cultural, economic and technological change." Prospective identities are shaped according to particular aspirations for the future, such as raising economic performance or installing new multicultural values. Through prospective identities, it is hoped that visions for the future can be stabilized. That is, students will carry these visions of the future into their own schooled mentalities and identities, learning how to see, think, and act in their own future lives in order to bring about the cultural, economic, and technological changes required.[2]

Neither retrospective nor prospective identities are naturally given. They are fabricated, invented, created in order to achieve the objectives of various kinds of authorities. Retrospective identities are usually associated with conservative cultural institutions and restorative ideology. In comparison, the invention of the curriculum of the future is the result of a diverse and heterogeneous network of authorities, actors, and organizations, all of which are seeking to project aspirations for the future of school. These aspirations are motivated by different objectives and visions. Some are economic, others more cultural, some concerned with technological change. Despite their differences, though, they do all promote new ideas, frameworks and objectives for the curriculum at a time of economic, cultural and technological change, and therefore they do all promote prospective identities.

Prototyping Identities

In the projects that constitute the curriculum of the future, new identities are being sculpted and "prototyped." That is to say, these programs are working to shape and make up new kinds

of identities for particular kinds of future aspirations. Some of
these prototypical identities are made very explicit in the various project documents. In the "new times" constructed as the
context for the New Basics intervention, a particular ideal of the
individual is created. As the project Web documentation states:

The New Basics categories capture various aspects of the person in the
world:

• the communicator—active and passive, persuading and being per-
suaded, entertaining and being entertained, expressing ideas and emotions in words, numbers and pictures, creating and performing
• the individual—physically and mentally, at work and at play and as a
meaning-maker
• the group member—in the family, in social groups, in government-
related groups and so on
• part of the physical world—of atoms and cells, electrons and chromo-
somes, animal, vegetable and mineral, observing, discovering, constructing and inventing.

An accompanying technical outline of the theoretical underpin-
nings for the New Basics links its approach to American critical
and "reconceptualist" models of the curriculum that, it claims,
"can be built by envisioning the kinds of life worlds and human
subjects that the education system wants to contribute to and
build." The person articulated in the project documents is a connected individual who, empowered by emerging network technologies, is able to move fluidly and fluently across "diverse
communities and complex cultures."[3]

There is something of a cosmopolitan identity imagined by
the New Basics: the individual at home anywhere in the world.
The objective for the New Basics, therefore, is with the remaking of certain sorts of people and cultures: the formation of a
prospective identity based on a particular interpretation of

technological, cultural, and economic change that have been projected into a series of curricular aspirations and objectives.

The Quest to Learn high school in New York, as well as its sister institution in Chicago, also embeds a strong prospective identity in its curriculum framework. The project texts state that learners are imagined as "sociotechnical engineers" with "network literacy" and the capacity for interdisciplinary "systems thinking," a "characteristic activity in both the media and science today."[4] These ways of knowing produce a prospective identity that can deal with complex technological change in futures that are going to be increasingly networked and require transdisciplinary expertise in the domains of media and science. The Web site for ChicagoQuest states very clearly its promotion of new student identities. It encourages "students to 'take on' the identities and behaviors of explorers, mathematicians, historians, writers, and evolutionary biologists as they work through a dynamic, challenge-based curriculum."[5] The prospective identity of Q2L is constructed for professional interdisciplinary innovation, though it also draws on young people's cultural experiences as participants in networked publics and global communities.

Learning Futures reimagines the future of school as a "base camp," a "hub that creates connections," and the prospective identity it fabricates is one that is able to move fluidly across formal educational institutions, intermediate institutions such as families and neighborhoods, and wider platforms and tools for learning across informal communities. Here we have a prospective identity that is itself constantly moving through a network of learning opportunities at school, home, community, and online. Learning Futures constructs a prospective identity that is concerned with the community but at the same time imagines students as "proto-professionals."

As the project documentation states, "Learning Futures schools are seeking to develop pedagogies which transform the identity of the learner from 'recipient of information' to thinking (and being) like a scientist, geographer, artist, entrepreneur."[6] Moreover, the project assumes that student engagement can be achieved through identifying and measuring "how students think, feel and act in school": it identifies these three elements as

- Thinking/Cognitive;
- Feeling/Emotional/Affective;
- Acting/Behavioural/Operative.[7]

The Learning Futures prospective identity is, therefore, a networked, proto-professional identity that thinks, feels, and acts in terms of cognitive, emotional, and behavioral categories: it is both entrepreneurial and psychological.

It is important to restate, however, that the prototype curriculum examples being examined in this report draw extensively on arguments and ideas from digital culture. As a consequence, we need to take into account the resources involved in the shaping and making up of young people's "digital identities."

Remixing Identities

Put simply, identity is the answer to questions such as "Who do I think I am?," "What do I think is my place in the world?," and "Who do I want to become?" With the proliferation of digital media and networked communications technologies in many aspects of public and private life, our identity questions today may be recast as "Who do I think I am, when I'm on Facebook?," "What do I think is my place in the world, in *World of*

Warcraft?," and "Who do I want to become, in my *Second Life*?" Do we possess one kind of identity in the analog world, and yet another in the digital world—a kind of "Identity 2.0"? Are identities possible when they have been detached from their bodies?

In such contexts, human identity is no longer thought about in terms of its unity, but in terms of a multiplicity, heterogeneity, and fragmentation of "cyberselves." The multiplicity of identity may be interpreted positively or negatively. The virtual dimensions of social networks allow for the fluidity and multiplicity of identity as an ongoing creative process of constructing "identities-in-action" and "work-in-progress," but also permit the construction of fractured, confused and "half-real" reflections of a person. The digital identities permitted by seeing ourselves as "plugged-in technobodies" are flexible and multiple and decentered in different roles in different settings at different times.[8]

The potential of "DIY media" is understood to "empower" young people in a do-it-yourself ethic of creative collaboration; production and participation. It puts the emphasis on the autonomy, agency, and creativity of users, or, as they have been fondly neologized, "pro-sumers" and "prod-users."[9] However, this pleasurable and playful multiplication of identities is also intensely political. In linking the requirement for lifelong learning to the DIY culture of the Web, self-editing and digital identity management become key lifelong skills as individuals are required to self-adjust or constantly update and upgrade their identities. Individuals are encouraged to become perpetually involved in optimizing themselves through DIY processes of accessorization and upgrading, enhancing their social reach through network extensions and ensuring the credibility, trustworthiness, and reputation of their profiles through constant processes of consumption. Put in these terms, identity is a performance that is social, political, economic, personal, and increasingly "*remixed and remixable.*"[10] The self-remixing DIY discourse stems from the promotion of a specific new kind of reflexive social identity that is active in practices of self-responsibility, self-shaping, and self-mastery.[11]

New hybrid identities are produced actively and reflexively as persons negotiate worlds that are both tangibly nearby and virtually dispersed. They are not given at birth but are the effect of constantly juggling multiple real-world and virtual identities, and working upon one's self as a personal project. Perhaps even more critically, it has been suggested that social network sites have reduced people to "multiple-choice identities" as a result of "locked-in" computer science templates.[12]

Looked at in this way, the kind of lifelong learning identities envisaged in various curriculum futures is the educational outgrowth of a DIY culture in which individuals are encouraged to see themselves and their lifestyles as constant creative projects. Identities are no longer given but need to be assembled like flat-pack furniture. In a DIY self-driven culture, learning become endless, lifelong, and lifewide across the entire life cycle, as individuals seek out new experiences and hence more learning. Learning is repositioned by digital media culture as a lifestyle choice rather than an institutionalized process of schooling.[13]

Specifically taking up such analyses, the Enquiring Minds project in the United Kingdom focused on the "making up" of the child. The emphasis on flexibility and adaptability in the face of new uncertainties creates a particular type of person, a reschooled identity characterized in the EM research as a "flexible child" who is "response-ready" and "response-able" and lives constantly in an "unfinished" state of self-innovation.[14]

As a result of the new kinds of remixable digital identities young people are constructing for themselves, reconfigured identities are to be required within the digitalized classrooms of the curriculum of the future. New kinds of identities are to be lashed up and reassembled alongside the refashioning of educational priorities, objectives, and strategies, and linked to new ways of thinking about such things as human communication, online consumption, and digital lifestyles. In the digital era the prospective identities and mentalities of the school child are to be "mashed up" from heterogeneous resources rather than defined through grand curricular narratives of the past.[15]

In the curriculum prototypes of Enquiring Minds, High Tech High, Quest to Learn, the New Basics, and so on, new identities are fabricated and promoted. Instead of "schooled identities," the projects promote a range of remixed and mashed-up identities, a kind of half-schooled/half-digital hybrid. These examples of centrifugal schooling represent a futuristic vision of education for the next century that suggests that networked individual identity building—rather than the acquisition of prepackaged "schooled identities" as embodied in formal curricula—is at the heart of educational modernization, innovation, and twenty-first-century reform. Centrifugal schooling extends the schooled identities of young people into an ongoing process of self-fulfillment and personal lifestyle creation that has now become the characteristic feature of lifelong learning in a modern consumer-media society.

The reconfiguration of formally schooled identities as fluid, self-fashioning digital learning identities also links young people more forcefully to changing working circumstances where the emphasis is on workers who can continually improve themselves, upskilling and retraining as changing job descriptions require. The enterprising selves, permanently unfinished projects, and interactive social identities of reflexive, self-adjusting, lifelong learners are essential as the human capital required by the knowledge economy as well as by the new global community.[16]

The digital learning identities promoted by centrifugal schooling are "cyborg" identities, hybrids of humans with information technologies, which connect the bodies and minds of young people into the disembodied and deterritorialized spaces of the Internet. The firm disciplinary identities of linear curricula are to be disassembled by the more centrifugal dynamics and fluidities of the digital age, and instead digital learning identities are to be reassembled in relation to lifelong learning, identity accessorization, enterprise, and notions of DIY identity construction. Digital learning identities are expressions of increasingly centrifugal selves and the mashed-up identities being constructed through the curriculum of the future are, then, reticulated cyborg identities.[17] The characteristics of cyborg identities are

• cyborg connectivity: being networked, connected, flexible, interactive, interdependent;

• projective competence: being psychologically self-competent, self-fashioning, self-upgrading, creative, and innovative, with the self as a personal project;

• prospective futures: being engaged in lifelong learning, problem solving.

Drawing on these clusters of cyborg connections, projective psychological competence, and prospective futures, it is possible to suggest that an idealized identity has been established across the range of curriculum prototypes examined. This identity is idealized as a lifelong networked learner with psychological eyes, or a DIY networked individual.

DIY Networked Individualism

In terms of lifelong learning, the DIY networked individualist prospective identity is constructed from a discourse of learning as an active and lifelong project. The curriculum may be understood as distributed across both formal and informal contexts, stretched lifelong and lifewide, with learning increasingly harmonized right across boundaries of educational space and pace. Rather than the educational spaces of schools with their classrooms and textbooks, learning happens in many formal and informal spaces, including home, school, community, and online spaces. And rather than the usual rhythmic pace of schooling according to timetables and the staged organization of curriculum, lifelong learning happens throughout the entire life cycle, in authentic contexts, just in time, and on-demand.

In terms of networked individualism, the prospective identity focuses on the personal projects of the individual. In the culture of networked individualism, the values, choices, interests, and projects of the individual are at the forefront. Individuals are now understood as having the capacity to be more active and knowing, to be participants in networked publics and creative audiences, with great potential for personal and cultural autonomy. This means that a culture of networked individualism can inspire project-oriented social movements and insurgent communities of practice based on the sharing of new values and the construction of new kinds of identities. But it can also lead to entrenchment in communities that affirm and ascribe identities, such as those provided by a seductive consumer media culture.

In the prototypical curriculum projects examined, the culture of networked individualism has been detectable in particular in the emphasis given to personal projects and portfolios.

The personal project has become a state of mind rather than simply an assignment. Students are encouraged to make projects for themselves that express their anxieties and their aspirations for the future, and they are encouraged to view their very own selves and their identities as ongoing DIY projects. The extended personal project embedded in many examples of the curriculum of the future is the ideal pedagogy for such a culture.

In terms of its psychological construction, the prospective identity associated with the curriculum of the future has been assembled according to psychological concepts (creativity, competence, cognition, affect, motivations, lifelong learning) rather than the academic and epistemological fields on which the subjects have been constructed historically by experts. The main sources of authority on the curriculum now are informed by an expertise derived from across the "psy complex" of disciplines. It is through psychological eyes and a "psy" gaze that the student of the future is being imagined by the reimagining of the curriculum of the future. Students are encouraged to think, feel, and act upon themselves psychologically as inner-focused persons with mental and emotional habits of mind and states of well-being that are to be sculpted in order to support an economy of creativity and innovation.

Moreover, in the interdisciplinary blending of psy discourses with computer science perspectives in the learning sciences, students are also being encouraged to see themselves as computer engineers see things. As a result, the prospective identity of the learner promoted by the curriculum of the future is shaped as a CompPsy hybrid. Of course, socially defined identities are never simply determined by external forces. But social identities can be promoted and sculpted in ways that position students in certain ways and encourage students to see themselves in their terms.

The expertise and authority of psychological eyes and computer science generate for students particular ways of viewing, thinking, feeling, and acting; not least for seeing, thinking about, and acting on themselves. In other words, students too are now being encouraged to identify with a particular style of thought, to think, see, and practice on themselves through particular types of concepts, key terms, references, explanations, arguments, and techniques. In line with the expertise of the CompPsy complex, students of the curriculum of the future are to be schooled to be self-activating, inner-focused, emotionally well, playful and creative, as well as experimental, innovative, transdisciplinary, entrepreneurial, and mentally flexible. Students are encouraged to see themselves as self-enterprising, autonomous, and creative individuals, taking charge of their own fates as a lifelong project. They are encouraged to attach themselves "prosthetically" via multiple networks, to "project" themselves through personal projects of the self, and to orient themselves "prospectively" toward the future. The curriculum of the future is not just a matter of defining content and official knowledge. It is about creating, sculpting, and finessing minds, mentalities, and identities, promoting style of thought about humans, or "mashing up" and "making up" the future of people.

8 Conclusion: An (Un)official Curriculum of the Future?

Changing ideas about the curriculum of the future show that what knowledge gets taught at school remains an important issue for debate. The curriculum acts as a microcosm of society, condensing what a society chooses to remember of its past, how it understands its present, and what it aspires and wants to project prospectively into the future. The curriculum prototypes analyzed here act as microcosms of where society wants to be heading in the future, and need to be examined not as socially independent or neutral bodies of content but in terms of their wider societal interdependence. A curriculum is not a disinterested, naturally predetermined or "given" body of knowledge. It is the result of an active process of engineering and tends to embody or mirror the political, economic, cultural, and social realities from which it emerges. Like many other complex things, a curriculum needs to be constructed, invented, assembled, or "made up." The creation of a curriculum is also a process of remaking society and remaking people.

The prototypical examples of new curriculum programs examined in the report show how the future of the curriculum

is now in the hands of a great many varied individuals and organizations, many of them from outside the mainstream education system. These agents and agencies collectively constitute a
new global curriculum design network with its own languages,
techniques, and motivations that are constructed upon the basis
of authority and expertise drawn from different professional disciplines, knowledge domains, and sets of political values. The
curriculum of the future is the subject and the product of a particular style of thought.

The "official knowledge" embedded in each of these prototypical curricula is, therefore, the result or effect of complex
ongoing processes, interpretations, negotiations, contests and
conflicts, and compromises and agreements that have constituted the formation of such a style of thought. That is to say,
these curriculum experiments are socially shaped. Every new
curriculum has its own social life. Each of them represents a juxtaposition and a synthesis of ideas, aspirations, and objectives
about such major societal issues as the future of the economy,
the impact of commercialization and privatization on public
education, changing notions of social expertise and authority,
the cultural patterns of communalism and individualism on the
Web, and the formation of young people's identities, mentalities, and minds.

Defining what counts as worthwhile knowledge for inclusion
in the curriculum of the future is not incidental to these issues:
it is constituted by the way these issues are addressed. In conclusion, let's review some of the main points from each of the
chapters. Together, these main points constitute the new style
of thought regarding the curriculum of the future: its key terms,
concepts, references, relations, arguments, explanations, and
the practical techniques deployed to modify or remake it.

Curriculum: The curriculum has, over the last couple of decades, been increasingly "harmonized" with a series of societal transformations linked to globalization and the political aspirations of nations to compete in a knowledge economy. The knowledge economy has become a preferred vision for the future of society, with the result that the curriculum has been put under intense pressure for reform. The consequence has been for reformers to put the emphasis on frameworks of skills, competences, "know-how" and other categories of "learning," and an evacuation of content, knowledge and "know-what" from the curriculum. Close analysis of these developments shows how they are formed from an uneasy alliance of economic arguments about the need to equip students with skills for digital labor and educational ideals drawn from a long history of progressivist and constructivist learning.

Networks: "Networks" have become part of a paradigmatic vocabulary for the centrifugal future of schooling. Networks are proposed as the ideal organizational form in a "smart" lateral world that now values mobility, fluidity, and dynamism over all rigidities and hierarchies. People now work through networks; they experience culture through networks; they engage with diverse publics through networks; and they may be exploited through their connections to different networks. Educational institutions and systems have come under sustained attack for their incapacity to keep up with the dynamism of a network based society, with the result that new innovations have focused on the development of more "open education" systems. The dominant emergent discourse is one of complexity, systems thinking, multiplicity, and dynamism. The Quest to Learn high school embodies how this discourse can be made into a productive curriculum framework. Other initiatives, however, utilize the Internet itself to distribute educational opportunities into a cloud culture of learning beyond the boundaries of school. Again, these approaches incorporate a progressivist legacy into a high-tech paradigm to create a networked neoprogressivist hybrid ideal of the curriculum of the future.

Economy: The knowledge economy makes new demands of schools, especially how students are schooled for work. However, the correspondence of the curriculum and work has been challenged by a new series of links and associations between schools and economic interests. Cultures of playful learning, an explosion of creativity, and commercialism combined now appear to promote new ways of thinking, feeling, and acting in schools that are linked to "reenchanted" economic or market values. Authority for the content of the curriculum has been assumed by a new mix of private-sector and public-sector objectives working together through "crossover" alliances. Rather than the state operating alone, curriculum development increasingly consist of a messy mix of governmental and nongovernmental organizations, private-sector and commercial companies, philanthropies, think tanks, and social enterprises. Its emphasis for the future of the curriculum, both in terms of governance and classroom practice, is increasingly on short-term, fast-time projects, all linked together through the "reenchanting" policy discourse of creativity.

Expertise: Partly as a result of new forms of crossover governance, new sources of professional and theoretical expertise and authority are now becoming involved in shaping the curriculum of the future. In everyday life, "little experts" are now increasingly taking the place of traditional authorities, particularly in the culture of the Internet. In the educational domain, such little experts take the guise of intellectual workers who take big abstract ideas and translate them into "vehicular ideas" that can be moved along quickly to get things done in classrooms. The curriculum of the future is partly the result of an explosion of expertise as new intellectual workers have begun to intervene in the education system from think tanks, corporate R&D labs, nonprofits, philanthropies, and academic departments alike. Their new expertise promotes new ways of knowing and acting in schools that derive from two main sources of authority: the psychological disciplines and computer sciences. In the psychological management of the curriculum of the future, great stress is put on learners' self-actualization and active self-responsibility. In addition, the blending of psychological disciplines with computer science disciplines in the transdisciplinary field of the learning sciences has created a new "CompPsy complex" that aims to make up a particular kind of self-competent, inner-focused individual whose emotional well-being is important for innovation and the future well-being of the economy. The result is that "psychotechnical schools" are now being encouraged to act upon the capacities and competencies of individuals in relation to perceived political and economic objectives.

Culture: In addition to overtly economic and political objectives, a curriculum also represents what society defines as "real culture" (or what real culture ought to be). The culture of the Internet is increasingly recognized as part of the real culture of the present and is therefore articulated as part of the cultural world to be represented in the curriculum of the future. The cultural patterns of the Internet can be roughly divided into communalist and individualist. Some examples of the curriculum of the future focus on communal patterns of cosmopolitanism and multiculturalism, while others are shaped and influenced by the ideal of "networked individualism" that understands individuals to be responsible for their own "projects." Curriculum projects also respond to the culture of branded consumerism and the growth of consumer-media culture as a seductive and informal curriculum of pleasurable lifestyle choices. In this culture of networked individualism, individuals are encouraged to participate constantly in active DIY projects of self-improvement and self-driven, lifelong learning.

Identities: In the networked world microcosmically represented in the curriculum of the future, new kinds of learner identities are promoted and shaped. In place of the retrospective "schooled" identities of students, young people are being sculpted and molded prospectively as lifelong learners with the competence and capacity to be flexible, self-adjusting, and self-enterprising in changing futures. Rather than linking learner identity to disciplinary knowledge, the curriculum of the future links identity to a hybridized learning landscape that cuts across formal and informal sites. The prospective identities of the curriculum of the future are lifelong networked individualists who see things through psychological eyes and comprehend them through computational thinking. Identities are increasingly considered to be a lifelong project that the individual constantly works upon. Instead of ready-made identities, all individuals are responsible for their own DIY identities, which they must manage fastidiously throughout their lives.

Toward an (Un)official Curriculum of the Future

As we have seen, the curriculum of the future is being socially shaped according to quite complex arguments about learning and knowledge, networks and systems, economics and expertise, and culture and identities. Together, these arguments, and the terms, concepts, references, and relations that underpin them, constitute an emerging style of thought regarding the curriculum. It is important to restate that these developments mostly remain prototypical and incomplete, and that much of the material covered is promotional rather than empirical. The final upshot of the analysis offered in this report is that the minds and mentalities of young people are subject to an emerging style of thought that seeks to shape, mold, and sculpt them as certain sorts of people in order to promote and enact a preferred vision of society. The extent to which things might happen as they have been imagined, promoted, and planned is a matter for further research on the ground.

The approach in this analysis has been critical, not out of aggressive critical militancy or a rush to judgment but out of an attempt to understand how changes being imagined in the content, form, and control of the curriculum are related to wider social, political, economic, and cultural matters. It is according to various social, political, economic, and cultural matters that any curriculum is made real and official; it does not just spring into existence ready-made but must always be assembled and made official as a representation of the past, a version of the present, and an aspiration for the future.

The visions for the future of society imagined by the various prototypical examples of the curriculum of the future all challenge the idea that a single, central, and official version of the curriculum is possible. Instead, they promote a much more centrifugal and decentralized vision of schooling. Centrifugal schooling, as the collective name given to the prototype curriculum projects, represents an emergent and unofficial vision of a curriculum of the future—a style of thought for the curriculum of the digital age. An empirical research program dedicated to examining and understanding the centrifugal organization of the unofficial curriculum of the future would further seek to explore these emerging features in concrete settings.

Centrifugal knowledge: Any curriculum represents a selection of knowledge, a construction of a reality to be passed from one generation to the next. Research on the curriculum of the future needs to dissect and analyze the knowledge contained in such programs. It needs to look at the structure of such knowledge and track its definite social relations. Do, for example, transdisciplinary approaches in the curriculum accurately track professions and generate appropriate (proto-)professional identities? What are the social conditions and contexts that have generated the knowledge that is to form the knowledge base of the curriculum? What communities of specialists have generated it? On what theories does it rest? Or is the knowledge included in the curriculum of the future divorced from the real contexts of knowledge production? Finally, if curriculum knowledge is to be defined according to more horizontal or "open source" ideals rather than by vertical hierarchy, what will give knowledge its authority and according to what theories and accounts will knowledge "count" as worthwhile?

Centrifugal authority: What are the specific sources of expertise and authority involved in promoting new curricular visions? The curriculum of the future involves a variety of individuals, organizations, cross-sectoral connections, and sources of expertise all being enrolled together to form new decentered amalgamations of authority. The state is no longer the central source of authority, and even when it continues to mandate and prescribe curriculum policies it does so indirectly through mediators, catalysts, fixers, and intellectual workers who bring new ideas, new theories, and new sources of expertise to the policy process. Further research on the curriculum of the future needs to trace the complex interorganizational and cross-sectoral processes, as well as the historical and political associations and networks involved in this amalgamation of curriculum authority

Centrifugal identities: In digital culture identity has been multiplied as individuals are permitted to perform their own selves in different digital environments. In the curriculum of the future, different identities and positions are promoted to students, with the idealized position being that of the self-actualizing, psychologically introspective networked individual and lifelong learner. This "cyborg" identity is prosthetically attached via networks, psychologically projected through projects of the self, and turned prospectively toward the future. Further curriculum research needs to examine through empirical analysis the ways in which students come to understand themselves and plan for their futures through different curricula. It needs to place identity in its necessary political context, as the human embodiment of political aspirations that have a preferred future vision of society and the remaking of learners' identities as their objective.

Notes

1 Introduction

1. The research for this report started life as part of a working group project on curriculum innovations in the digital age generously supported by the Digital Media and Learning Research Hub working groups competition 2010 with funding from the John D. and Catherine T. MacArthur Foundation. Thanks are due in particular to the working group participants, Ola Erstad, Øystein Gilje, Jen Groff, John Morgan, Sarah Payton, Rebecca Rufo-Tepper, and Arana Shapiro, whose insider experience of curriculum innovation helped shape the research.

2. Seminal research on curriculum and education reform from either side of the Atlantic includes Michael W. Apple, *Official Knowledge: Democratic Education in a Conservative Age*, 2nd ed. (New York: Routledge, 2000); Stanley Aronowitz and Henry A. Giroux, *Education Still under Siege*, 2nd ed. (Westport, CT: Bergin and Garvey, 1993); Stephen J. Ball, *Education Reform: A Critical and Post-structuralist Approach* (Buckingham, UK: Open University Press, 1994); David C. Berliner and Bruce J. Biddle, *The Manufactured Crisis: Myths, Fraud, and the Attack on America's Public Schools* (Reading: MA: Addison-Wesley, 1995); Basil Bernstein, *Pedagogy, Symbolic Control and Identity: Theory, Research, Critique*, rev. ed. (Oxford: Rowman and Littlefield, 2000); Ivor F. Goodson, *Learning, Curriculum*

and Life Politics (New York: Routledge, 2005); Linda McNeil, *Contradictions of School Reform: Educational Costs of Standardized Testing* (New York: Routledge, 2000); William F. Pinar, *What Is Curriculum Theory?* (Mahwah, NJ: Erlbaum, 2004); Michael F. D. Young, *Bringing Knowledge Back In: From Social Constructivism to Social Realism in the Sociology of Education* (New York: Routledge, 2008).

3. The 1983 text of *A Nation at Risk: The Imperative of Educational Reform* is archived at http://www2.ed.gov/pubs/NatAtRisk/index.html.

4. Thomas Osborne and Nikolas Rose, "Governing Cities: Notes on the Spatialisation of Virtue," *Environment and Planning D: Society and Space* 17 (1999): 737–760; Nikolas Rose, *The Politics of Life Itself: Biomedicine, Power, and Subjectivity in the Twenty-First Century* (Princeton, NJ: Princeton University Press, 2007).

5. Wiebe Bijker and John Law, *Shaping Technology/Building Society: Studies in Sociotechnical Change* (Cambridge, MA: MIT Press, 1992); Noel Gough, "Voicing Curriculum Visions," in *Curriculum Visions*, ed. William E. Doll and Noel Gough (New York: Peter Lang, 2002), 1–22; Ciaran Sugrue, ed., *The Future of Educational Change: International Perspectives* (New York: Routledge, 2009).

6. Alex Molnar, *School Commercialism: From Democratic Ideal to Market Commodity* (New York: Routledge, 2005).

7. Michael W. Apple, *Educating the "Right" Way: Markets, Standards, God, and Inequality*, 2nd ed. (New York: Routledge, 2006); Stephen J. Ball, *Education plc. Understanding Private Sector Participation in Public Sector Education* (New York: Routledge, 2007); Joel Spring, *Globalization of Education: An Introduction* (New York: Routledge, 2009).

8. David Buckingham, *The Material Child: Growing Up in Consumer Culture* (Cambridge, UK: Polity, 2011).

9. Peter Miller and Nikolas Rose, *Governing the Present: Administering Economic, Social and Personal Life* (Cambridge, UK: Polity, 2008).

10. Stephen J. Ball, *The Education Debate* (Bristol, UK: Policy Press, 2008); Stephen J. Ball, *Global Education Inc.: New Policy Networks and the*

Neo-liberal Imaginary (New York: Routledge, 2012); Faizal Rizvi and Bob Lingard, *Globalizing Education Policy* (New York: Routledge, 2010).

11. Ben Williamson, "Centrifugal Schooling: Third Sector Policy Networks and the Reassembling of Curriculum Policy in England," *Journal of Education Policy* (2012), doi: 10.1080/02680939.2011.653405.

12. Constance M. Yowell, "Connected Learning: Designed to Mine the New, Social, Digital Domain," *DMLcentral.net*, March 1, 2012, http://www.dmlcentral.net/blog/constance-m-yowell-phd/connected-learning-designed-mine-new-social-digital-domain.

13. Zygmunt Bauman, *Liquid Times* (Cambridge, UK: Polity, 2007).

14. Manuel Castells, *Communication Power* (New York: Oxford University Press, 2009); Henry Jenkins, *Convergence Culture: Where Old and New Media Collide* (New York: New York University Press, 2006).

15. Juha Suoranta and Tere Vaden, *WikiWorld* (London, UK: Pluto Press, 2010).

16. Cathy N. Davidson and David T. Goldberg, *The Future of Learning Institutions in a Digital Age* (Cambridge, MA: MIT Press, 2009).

17. Trebor Scholz, ed., *Learning through Digital Media: Experiments in Technology and Pedagogy* (New York: Institute for Distributed Creativity, 2011).

18. Michael F. D. Young, *The Curriculum of the Future: From the 'New Sociology of Education' to a Critical Theory of Learning* (London: Falmer Press, 1998).

19. Sharon Gewirtz and Alan Cribb, *Understanding Education: Sociological Perspectives* (Cambridge, UK: Polity, 2009).

20. This framework is influenced by the approaches of Michel Foucault and Bruno Latour as they have been redeployed in educational research and other related sociological studies.

21. Stephen J. Ball, *Education Reform: A Critical and Post-Structural Approach* (Buckingham, UK: Open University Press, 1994); Stephen J.

Ball, Meg Maguire and Annette Braun, *How Schools Do Policy: Policy Enactments in Secondary Schools* (New York: Routledge, 2012).

22. Tara Fenwick and Richard Edwards, *Actor-Network Theory in Education* (New York: Routledge, 2010); Tara Fenwick, Richard Edwards, and Peter Sawchuk, *Emerging Approaches to Educational Research: Tracing the Sociomaterial* (New York: Routledge, 2012).

23. http://www.enquiringminds.org.uk. I was a researcher on this project throughout its duration.

24. http://www.hightechhigh.org/; http://gse.hightechhigh.org/.

25. http://www.innovationunit.org/our-services/projects/learning-futures-increasing-meaningful-student-engagement

26. http://education.qld.gov.au/corporate/newbasics/.

27. http://www.rsaopeningminds.org.uk/.

28. http://www.q2l.org.

29. http://www.chicagoquest.org.

30. http://www.p21.org.

31. http://www.wholeeducation.org.

2 Curriculum Change and the Future of Official Knowledge

1. Pinar, *What Is Curriculum Theory?*; William F. Pinar, William M. Reynolds, Patrick Slattery, and Peter M. Taubman, *Understanding Curriculum: An Introduction to the Study of Historical and Contemporary Curriculum Discourses* (New York: Peter Lang, 1995).

2. Apple, *Official Knowledge*.

3. Dave Scott, *Critical Essays on Major Curriculum Theorists* (New York: Routledge, 2008).

4. Berliner and Biddle, *The Manufactured Crisis*.

5. Useful critical essays on the knowledge economy from American and British perspectives are provided in Hugh Lauder, Michael Young, Harry Daniels, Maria Balarin and John Lowe, eds., *Educating for the Knowledge Economy: Critical Perspectives* (New York: Routledge, 2012).

6. Michael F. D. Young, *The Curriculum of the Future: From the 'New Sociology of Education' to a Critical Theory of Learning* (London: Falmer Press, 1998).

7. Aronowitz and Giroux, *Education Still under Siege*; Ball, *Education Reform*.

8. Pinar, *Curriculum Theory?*; Anthony Seldon, *An End to Factory Schools: A Manifesto for Education 2010–2020* (London: Centre for Policy Studies, 2010).

9. Lois Weiner, "Schooling to Work," in *Post-Work: The Wages of Cybernation*, ed. Stanley Aronowitz and Jonathan Cutler (New York: Routledge, 1998), 185–201.

10. Ball, *Education Reform*, 46.

11. James Paul Gee, *Situated Learning and Literacy: A Critique of Traditional Schooling* (New York: Routledge, 2004).

12. Philip Brown, Hugh Lauder and David Ashton, *The Global Auction: The Broken Promise of Education, Jobs, and Incomes* (New York: Oxford University Press, 2011).

13. International Society of the Learning Sciences, Overview (2004), http://www.isls.org/about.html.

14. Gert Biesta, *Beyond Learning: Democratic Education for a Human Future* (Boulder, CO: Paradigm, 2006).

15. Margaret Weigel, Carrie James, and Howard Gardner, "Learning: Peering Backward and Looking Forward in the Digital Era," *International Journal of Learning and Media* 1, no. 1 (2009): 1–18.

16. Michael F. D. Young, "Education, Globalisation and the 'Voice of Knowledge,'" in *Educating for the Knowledge Economy: Critical Perspectives*,

ed. Hugh Lauder, Michael Young, Harry Daniels, Maria Balarin and John Lowe (New York: Routledge, 2012), 139–151.

17. Hank Bromley and Michael W. Apple, eds., *Education/Technology/ Power: Educational Computing as a Social Practice* (Albany: State University of New York Press, 1998).

18. See *Demos Quarterly* 1 (Winter 1993), http://www.demos.co.uk/files/ openingminds.pdf?1240939425.

19. John Morgan, "Enquiring Minds: A Radical Curriculum Project?" *Forum* 53, no. 2 (2011): 261–272.

20. Sara Candy, "Opening Minds: A Curriculum for the 21st Century," *Forum* 53, no. 2: 286.

21. For an elaboration of the distinction between performance and competence modes, see Bernstein, *Pedagogy, Symbolic Control and Identity*.

22. Partnership for 21st Century Skills, *21st Century Readiness for Every Student: A Policymaker's Guide* (Tucson, AZ: Partnership for 21st Century Skills, 2011), http://www.p21.org/storage/documents/policymakers guide_final.pdf.

23. http://www. p21.org. P21 also offers a "one-stop shop" of information, resources, and community tools, Route 21, which features video downloads and an extensive resources library, http://route21.p21.org.

24. David Hartley, *Re-schooling Society* (London: RoutledgeFalmer, 1997); Young, *The Curriculum of the Future*.

25. Lauder et al., *Educating for the Knowledge Economy?*.

26. Young, *Bringing Knowledge Back*.

27. Brown, Lauder, and Ashton, *Global Auction*.

3   Networks, Decentered Systems, and Open Educational Futures

1. Luc Boltanski and Eve Chiapello, *The New Spirit of Capitalism* (London: Verso); Manuel Castells, *The Rise of the Network Society* (Oxford,

UK: Blackwell, 1996); Helen McCarthy, Paul Miller, and Peter Skidmore, eds., *Network Logic: Who Governs in an Interconnected World?* (London, UK: Demos, 2004); Nigel Thrift, *Knowing Capitalism* (Thousand Oaks, CA: Sage).

2. Johnny Ryan, *A History of the Internet and the Digital Future* (London: Reaktion, 2010).

3. Slavoj Zizek, *Violence* (London: Profile, 2008); Bauman, *Liquid Times.*

4. danah boyd, "Social Network Sites as Networked Publics: Affordances, Dynamics, and Implications," in *A Networked Self: Identity, Community, and Culture on Social Network Sites*, ed. Zizi Papacharissi (New York: Routledge, 2011), 39–58; Ito et al., *Hanging Out.*

5. Davidson and Goldberg, *The Future of Learning Institutions*; Alan Liu, *The Laws of Cool: Knowledge Work and the Culture of Information* (Chicago: University of Chicago Press, 2004); Scholz, *Learning through Digital Media.*

6. Michael A. Peters, "'Openness' and the Global Knowledge Commons: An Emerging Mode of Social Production for Education and Science," in *Educating for the Knowledge Economy?: Critical Perspectives*, ed. Hugh Lauder, Michael Young, Harry Daniels, Maria Balarin, and John Lowe, 66–76 (New York: Routledge, 2012).

7. Katie Salen, Rob Torres, Laura Wolozin, Rebecca Rufo-Tepper, and Arana Shapiro, *Quest to Learn: Developing the School for Digital Kids* (Cambridge, MA: MIT Press, 2011).

8. James Paul Gee, *What Video Games Have to Teach Us about Learning and Literacy* (New York: Palgrave MacMillan, 2003); Katie Salen, ed., *The Ecology of Games: Connecting Youth, Games, and Learning* (Cambridge, MA: MIT Press, 2008).

9. Salen et al., *Quest to Learn*, 32–33.

10. David Hartley, "The Instumentalization of the Expressive," in *Schooling, Society and Curriculum*, ed. Alex Moore (New York: Routledge, 2006), 60–70.

11. William E. Doll, "Complexity and the Culture of Curriculum," *Educational Philosophy and Theory* 40, no. 1 (2008): 190–212.

12. Fenwick, Edwards and Sawchuk, *Emerging Educational Research.*

13. All references are from the New Basics project Web site: http://education.qld.gov.au/corporate/newbasics/.

14. Keri Facer and Hannah Green, "Curriculum 2.0: Educating the Digital Generation," in *Unlocking Innovation: Why Citizens Hold the Key to Public Service Reform*, ed. Simon Parker (London, UK: Demos, 2007), 47–58.

15. http://schoolofeverything.com. For a similar initiative in America, see http://p2pu.org.

16. Suoranta and Vadén, *WikiWorld.*

17. Hartley, *Re-schooling Society*, 155.

18. Frank Webster, *Theories of the Information Society*, 3rd ed. (New York: Routledge, 2006).

19. Critical accounts of networks in education include Jo Frankham, "Network Utopias and Alternative Entanglements for Educational Research and Practice," *Journal of Education Policy* 21, no. 6 (2006): 661–677; Jorge Avila de Lima, "Thinking More Deeply about Networks in Education," *Journal of Educational Change* 11 (2010): 1–21; Kathleen Ferguson and Terri Seddon, "Decenterd Education: Suggestions for Framing a Socio-spatial Research Agenda," *Critical Studies in Education* 48, no. 1 (2007): 111–129.

20. Ferguson and Seddon, "Decenterd Education," 117–118.

21. Jaron Lanier, *You Are Not a Gadget* (London, UK: Penguin, 2010).

22. Castells, *Communication Power*, 413.

23. Jaron Lanier, "Does the Digital Classroom Enfeeble the Mind?" *New York Times*, September 16, 2010.

24. Pinar, *Curriculum Theory*, 8. Also on the "miseducation" of the Internet, see Ellen Seiter, *The Internet Playground: Children's Access, Entertain-*

ment and Mis-Education (New York: Peter Lang, 2005), and Molnar, *School Commercialism.*

4 Creative Schooling and the Crossover Future of the Economy

1. David Held, *Cosmopolitanism: Ideals and Realities* (Cambridge, UK: Polity, 2010); Bob Jessop, *The Future of the Capitalist State* (Cambridge, UK: Polity, 2002).

2. For a useful recent synthesis of the contesting positions in this debate, see David Baker, "The Educational Transformation of Work: A Synthesis," in Hugh Lauder et al., *Educating for the Knowledge Economy?: Critical Perspectives*, ed. Hugh Lauder et al. (New York: Routledge, 2012), 97–113.

3. http://www.hightechhigh.org/about/index.php.

4. Representative arguments can be found in Liu, *The Laws of Cool*; Jim McGuigan, *Cool Capitalism* (London: Pluto Press, 2009); Jamie Peck, *Constructions of Neoliberal Reason* (New York: Oxford University Press, 2010).

5. Michael Hardt, "Affective Labour," *Generation Online* (2008), http://www.generation-online.org/p/fp_affectivelabour.htm.

6. Gilles Deleuze, "Postscript on the Societies of Control," *October* 59 (1992): 3–7.

7. Thomas S. D. Osborne, "Against 'Creativity': A Philistine Rant," *Economy and Society* 32, no. 4 (2003): 507–525.

8. David Hartley, "Marketing and the 'Re-enchantment' of School Management," *British Journal of Sociology of Education* 20, no. 3 (1999): 309–323; Ben Williamson, "Effective" or "Affective" Schools? Technological and Emotional Discourses of Educational Change, *Discourse: Studies in the Cultural Politics of Education* 33, no. 3 (2012): 425–441; Nigel Wright, "Leadership, Bastard Leadership, and Managerialism," *Educational Management, Administration and Leadership* 29, no. 3 (2001): 275–290.

9. Hannah Green and Celia Hannon, *Their Space: Education for a Digital Generation* (London: Demos, 2007), 23.

10. Pat Kane, *The Play Ethic: A Manifesto for a Different Way of Living* (London: Pan MacMillan, 2004); Julian Kucklich, "Precarious Playbor: Modders and the Digital Games Industry," *The Fibreculture Journal* 5 (2005), http://five.fibreculturejournal.org/fcj-025-precarious-playbour-modders-and-the-digital-games-industry/; Trebor Scholz, "Facebook as Playground and Factory," in *Facebook and Philosophy*, ed. D. E. Wittkower (Chicago: Open Court/Carus Publishing, 2010), 241–252.

11. Alec Patton, *Work That Matters: The Teacher's Guide to Project-Based Learning* (London: Paul Hamlyn Foundation, 2012).

12. Important studies are Apple, *Official Knowledge*; Ball, *Education plc*; Ball, *Global Education Inc.*; Buckingham, *Material Child*; Molnar, *School Commercialism*; Seiter, *Internet Playground*; Joel Spring, *Globalization of Education*.

13. Benjamin R. Barber, *Consumed: How Markets Corrupt Children, Infantilize Adults, and Swallow Citizens Whole* (New York: W. W. Norton and Co., 2007).

14. Rizvi and Lingard, *Globalizing Education Policy*.

15. Ball, *The Education Debate*.

16. In the United Kingdom, for example, the think tank Demos has produced multiple publications linking educational innovation to high-tech R&D.

17. Fenwick and Edwards, *Actor-Network Theory in Education*.

18. http://www2.futurelab.org.uk.

19. http://www.enquiringminds.org.uk.

20. http://www.wholeeducation.org.

21. Salen et al., *Quest to Learn*.

22. http://www.newvisions.org.

23. Fenwick, Edwards, and Sawchuk, *Emerging Education Research.*

24. Stephen J. Ball and Sonia Exley, "Making Policy with 'Good Ideas': Policy Networks and the 'Intellectuals' of New Labour," *Journal of Education Policy* 25, no. 2 (2010): 151–169; David Hartley, "Rhetorics of Regulation in Education after the Global Economic Crisis," *Journal of Education Policy* 25, no. 6 (2010): 785–791; Ben Williamson, "Centrifugal Schooling: Third Sector Policy Networks and the Reassembling of Curriculum Policy in England," *Journal of Education Policy* (2012), doi: 10:1080/02680939.2011.653405.

25. On the distinctions between visible and invisible practices, see Basil Bernstein, "Social Class and Pedagogic Practice," in *Sociology of Education*, ed. Stephen J. Ball (New York: Routledge, 2004), 196–217.

5 Psychotechnical Schools and the Future of Educational Expertise

1. Michel Foucault, *Security, Territory, Population: Lectures at the College de France, 1977–1978*, trans. Graham Burchell (New York: Palgrave Macmillan, 2007).

2. Mitchell Dean, *Governmentality: Power and Rule in Modern Society*, 2nd ed. (Thousand Oaks, CA: Sage, 2010); Kenneth Hultqvist and Gunilla Dahlberg, eds., *Governing the Child in the New Millennium* (New York: RoutledgeFalmer, 2001); Nikolas Rose, *Governing the Soul: The Shaping of the Private Self*, 2nd ed. (New York: Free Association Books, 1999).

3. Miller and Rose, *Governing the Present; Nikolas Rose, Powers of Freedom: Reframing Political Thought* (Cambridge, UK: Cambridge University Press, 1999).

4. Gregor McLennan, "Traveling with Vehicular Ideas: The Case of the Third Way," *Economy and Society* 33, no. 4 (2004): 484–499; Thomas S. D. Osborne, "On Mediators: Intellectuals and the Ideas Trade in the Knowledge Society," *Economy and Society* 33, no. 4 (2004): 430–447.

5. Ball and Exley, *Good Ideas*; Williamson, *Centrifugal Schooling*.

6. Learning Futures, "Engaging Schools" (London, UK: Innovation Unit, 2010); Patton, *Work That Matters*.

7. Rose, *Governing the Soul*.

8. Lynn Fendler, "Educating Flexible Souls: The Construction of Subjectivity through Developmentality and Interaction," in *Governing the Child in the New Millennium*, ed. Kenneth Hultqvist and Gunilla Dahlberg (New York: Routledge), 132–133.

9. Bernstein, *Pedagogy, Symbolic Control and Identity*. For a learning theory perspective on competence, also see Weigel, James, and Gardner, "Learning."

10. Osborne, "Against 'Creativity,'" 508–509.

11. Kimberly Seltzer and Tom Bentley, *The Creative Age: Knowledge and Skills for the New Economy* (Buckingham, UK: Demos, 1999), 10.

12. Rob Pope, *Creativity: Theory, History, Practice* (New York: Routledge, 2005), 27.

13. http://www.wholeeducation.org.uk.

14. Thomas S. Popkewitz, "Numbers in Grids of Intelligibility: Making Sense of How Educational Truth Is Told," in *Educating for the Knowledge Economy?: Critical Perspectives*, ed. Hugh Lauder et al. (New York: Routledge, 2012), 177.

15. Rose, *Governing the Soul*.

16. http://appsforgood.org/course/.

17. Bill Lucas and Guy Claxton, *Wider Skills for Learning* (London: NESTA, 2009), 4.

18. Keri Facer and Jess Pykett, *Developing and Accrediting Personal Skills and Competences* (Bristol, UK: Futurelab, 2007); Nicola Bacon, Marcia Brophy, Nina Mguni, Geoff Mulgan, and Anna Shandro, *The State of Happiness: Can Public Policy Shape People's Wellbeing and Resilience?* (London, UK: Young Foundation, 2010).

19. Miller and Rose, *Governing the Present*.

20. Brown, Lauder and Ashton, *Global Auction*.

21. Rose, *Governing the Soul*, 116.

22. Neil Selwyn, "Looking beyond Learning: Notes toward the Critical Study of Educational Technology," *Journal of Computer Assisted Learning* 26 (2010): 65–72.

23. Torin Monahan, *Globalization, Technological Change, and Public Education* (New York: Routledge, 2005).

24. Thomas S. Popkewitz, *Cosmopolitanism and the Age of School Reform: Science, Education, and Making Society by Making the Child* (New York: Routledge, 2008).

25. Popkewitz, "Numbers in Grids of Intelligibility."

6 Globalizing Cultures of Lifelong Learning

1. Jenkins, *Convergence Culture*.

2. Castells, *Communication Power*.

3. Castells, *Communication Power*, 118–125.

4. Castells, *Communication Power*, 125.

5. http://education.qld.gov.au/corporate/newbasics/.

6. Rizvi and Lingard, *Globalizing Education Policy*, discuss the New Basics from the perspective of globalization and education policy.

7. Held, *Cosmopolitanism*.

8. John Morgan, Ben Williamson, Tash Lee, and Keri Facer, *Enquiring Minds: A Guide* (Bristol, UK: Futurelab, 2007), 14–15.

9. Goodson, *Learning, Curriculum and Life Politics*.

10. Morgan et al., *Enquiring Minds Guide*, 29.

11. Morgan et al., *Enquiring Minds Guide*, 24.

12. Facer and Pykett, *Developing and Accrediting*.

138   **Notes**

13. Patton, *Work That Matters*.

14. Peter Jarvis, "Globalisation, the Learning Society and Comparative Education," in *Sociology of Education*, ed. Stephen J. Ball (New York: RoutledgeFalmer, 2004), 72–86.

15. Gee, *Situated Learning*.

16. Xavier Bonal and Xavier Rambla, "Captured by the Totally Pedagogised Society: Teachers and Teaching in the Knowledge Economy," *Globalisation, Societies and Education* 1, no. 2 (2003): 169–184.

17. Buckingham, *Material Child*; Jane Kenway and Elizabeth Bullen, *Consuming Children: Entertainment-Education-Advertising* (Maidenhead, UK: Open University Press, 2001); Ellen Seiter, *Sold Separately: Parents and Children in Consumer Culture* (Bloomington: Indiana University Press, 1993).

18. Molnar, *School Commercialism*, 81 and 44–45.

19. Daniel T. Cook, "The Missing Child in Consumption Theory," *Journal of Consumer Culture* 8, no. 2 (2008): 219–243.

20. Lydia Martens, "Learning to Consume—Consuming to Learn: Children at the Interface between Consumption and Education," *British Journal of Sociology of Education* 26, no. 3 (2005): 343–357; Robin Usher, "Consuming Learning," in *Critical Pedagogies of Consumption: Living and Learning in the Shadow of the "Shopocalypse,"* ed. Jennifer A. Sandlin and Peter McLaren (New York: Routledge, 2009), 42.

21. Ball, *Education Debate*; Rose, *Powers of Freedom*.

22. Neil Selwyn, *Schools and Schooling in the Digital Age: A Critical Analysis* (New York: Routledge, 2011).

23. John Morgan and Ben Williamson, *Enquiring Minds: Schools, Knowledge and Educational Change* (Bristol, UK: Futurelab, 2008).

7 Making Up DIY Learner Identities

1. Popkewitz, "Numbers in Grids of Intelligibility."

2. A theory of retrospective and prospective learner identities is supplied by Bernstein, Pedagogy, Symbolic Control and Identity.

3. http://education.qld.gov.au/corporate/newbasics/.

4. Salen et al., *Quest to Learn.*

5. http://www.chicagoquest.org.

6. Learning Futures, "Engaging Schools."

7. David Price, "Learning Futures: Rebuilding Curriculum and Pedagogy around Student Engagement," *Forum* 53, no. 2 (2011): 273–284.

8. Nancy K. Baym, *Personal Connections in the Digital Age* (Cambridge, UK: Polity, 2010); Sherry Turkle, *Life on the Screen: Identity in the Age of the Internet* (London, UK: Phoenix, 1996); Sandra Weber and Claudia Mitchell, "Imagining, Keyboarding, and Posting Identities: Young People and New Media Technologies," in *Youth, Identity, and Digital Media*, ed. David Buckingham (Cambridge, MA: MIT Press, 2008), 25–48.

9. Axel Bruns, *Blogs, Wikipedia, Second Life and Beyond: From Production to Produsage* (New York: Peter Lang, 2008); Michel Knobel and Colin Lanskshear, eds., *DIY Media: Creating, Sharing and Learning with New Technologies* (New York: Peter Lang, 2010).

10. Zizzi Papacharissi, ed., *A Networked Self: Identity, Community, and Culture on Social Network Sites* (New York: Routledge, 2011).

11. Rose, *Governing the Soul*; Rose, *Powers of Freedom.*

12. Lanier, *You Are Not a Gadget.*

13. Zygmunt Bauman, *The Art of Life* (Cambridge, UK: Polity, 2008); Ulrich Beck and Elisabeth Beck-Gernsheimm, *Individualization: Institutionalized Individualism and Its Social and Political Consequences*, trans. Patrick Camiller (Thousand Oaks, CA: Sage, 2002).

14. Morgan and Williamson, *Enquiring Minds.*

15. Fenwick and Edwards, *Actor-Network Theory in Education*, 168.

16. Miller and Rose, *Governing the Present.*